PACEMAKER® PRACTICAL ARITHMETIC SERIES

Working Makes Sense

Charles H. Kahn

J. Bradley Hanna

Fearon/Janus/Quercus
Belmont, California

Simon & Schuster Education Group

PACEMAKER® PRACTICAL ARITHMETIC SERIES

Buying with Sense

Money Makes Sense

Using Dollars and Sense

Working Makes Sense

Contents

Cover Design: Joe C. Shines
Illustrators: Sam Masami Daijogo and Diana Thewlis

Copyright © 1989 by Fearon/Janus/Quercus, a division of Simon & Schuster
Education Group, 500 Harbor Boulevard, Belmont, CA 94002. All rights reserved.
No part of this book may be reproduced by any means, transmitted, or translated
into a machine language without written permission from the publisher.

ISBN 0–8224–7493–X

Printed in the United States of America

    4 5 6 7 8 9 10
CO

Working Makes Sense

Why do you think people work? "To make money," you might answer. And you would be right. Most people must work to earn a living.

By working, you earn the money you need to live on your own. You earn money for things you need, like food and clothes. You earn money to pay your bills. And you earn money to buy the things you want. Making money is a good reason for working. Making money is one reason that working makes sense.

But there is more to working than just making money. There are other good reasons for working, too. Work can be hard. At the end of the day, you may feel tired. But you can also feel pleased with yourself. You will feel good inside knowing that you have done your work well. No matter what kind of work you do, you have a right to feel important. Doing a good job is one of the most important things that anyone can do. It is another reason working makes sense.

When you work, you use money almost every day. And when you use your money, you need skills in addition, subtraction, multiplication, and division.

This book will help you learn the skills you need to use the money you earn. You will learn to use decimals. You will also learn about percents and fractions. You will learn about savings and checking accounts, too. These new skills will help you to use wisely the money you earn. And by using your earnings wisely, you will see that working makes sense.

Pretest I

Answer these questions about working.

1. Suppose you work from 8:00 A.M. to 4:30 P.M. each day. You take one hour off for lunch. How many hours do you work each day? _____

2. Suppose you earn $4.35 an hour. If you work 8 hours, how much money will you earn? _____

   If you work 40 hours, how much money will you earn? _____

3. Suppose you earn $5.20 an hour. How much money will you earn for working 6½ hours? _____

4. When you work more than 40 hours a week, you are paid time and a half. If you earn $4.50 an hour for regular time, how much will you earn for one hour of overtime? _____

5. You earn $850 in one month. If you save ⅛ of it, how much money will you save? _____

6. You have $65.00 in your savings account. If your money earns 5% interest each year, how much money will you have altogether after one year? _____

7. Every week you drive 98 miles to and from work. Your car uses 3.5 gallons of gas for these trips. How many miles per gallon does the car get? _____

8. Suppose you earn $866.00 a month. You are given a 15% raise. How much will you make a month now? _____

Work these problems on checking accounts.

9. Write out this check to Fiesta Foods for $78.16. Use today's date and sign your name.

	NUMBER 209
	01-05
	———
	638
	_____ 19 ____
PAY TO THE ORDER OF _____	$ _____
_____	DOLLARS
nonnegotiable	
CITIZEN'S BANK	
Barrington, Oklahoma	_____
⑆0638⑆ ⑈0105⑈ 34 1297 2⑆	

10. Complete this checkbook register. Figure the new balance for each check written and each deposit. *Subtract* the checks and *add* the deposits.

CHECK NO.	DATE	CHECKS ISSUED TO OR DESCRIPTION OF DEPOSIT	AMOUNT OF CHECK		√	AMOUNT OF DEPOSIT		BALANCE	
								416	09
209	5/10	Fiesta Foods (groceries)	78	16					
210	5/12	Dr. Frank (dentist bill)	45	00					
211	5/12	Gant's Dept. Store (clothes)	54	98					
	5/15	deposit				447	29		
212	5/18	Super Oil Co. (gas)	67	76					
213	5/20	Franklin Garage (oil change)	17	84					
214	5/21	Ma Bell (telephone)	25	19					
215	5/21	Entropy Power Co. (electric bill)	28	30					
216	5/24	West Bank (car payment)	225	00					
	5/31	deposit				387	29		
217	5/31	Hillcrest Apartments (rent)	550	00					
218	5/31	United Charge-it-All (credit card)	39	63					

Pretest II

Find the answers to these arithmetic problems.

1.
```
   67
   79
   86
+ 42
```

2.
```
  $3.95
- 1.82
```

3.
```
$26.46
×    7
```

4.
```
  $72.95
- 31.64
```

5. 4⟌256

6.
```
$8.31
 6.50
+  38
```

7. 6.4⟌72

8.
```
  $314.06
-  71.48
```

9.
```
$128.45
×   .06
```

10.
```
   $379.42
- 226.37
```

11. 12⟌104.76

12.
```
  $82.84
- 69.32
```

13.
```
$87.30
×    49
```

14.
```
  $6.61
  8.37
  4.55
+ 3.19
```

15.
```
$6.97
×   38
```

16. $3\overline{)\$8.25}$

17. $\$714.00$
$\underline{\times\ .31}$

18. $\$841.93$
$\underline{-638.67}$

19. $\$983.42$
766.11
$\underline{+\ 353.64}$

20. $1.5\overline{)12.9}$

21. Find 7% of $12.00. _____

22. Find 10% of $318.00. _____

23. Find 14% of $33.25. _____

24. Find 50% of $426.36. _____

25. Find 75% of $23.04. _____

26. Find 1/2 of $13.26. _____

27. Find 2/3 of $114.72. _____

28. Find 3/8 of 136 lbs. _____

29. Find 1/4 of 36.4 lbs. _____

30. Find 3/4 of $51.96. _____

Carla Gets a Job

Carla Williams was looking for a part-time job. She heard that Runaway Sports Shop needed someone to work evenings and weekends. Carla talked to the store's owner, Mr. Swift. He told her about the pay and the hours for the job. The job sounded good to Carla so she decided to work for Mr. Swift.

The day Carla started her job, Mr. Swift showed her how to use the time clock. "Many stores, offices, and factories have time clocks," Mr. Swift said. "A time clock makes it easy to keep a record of the hours you work. You stamp your time card in the clock when you start work. You stamp it again when you stop work. The time card shows the hours you worked."

Here is Carla's time card for the first week she worked. From Monday through Friday, Carla worked evenings from 5:00 P.M. to 9:00 P.M. On Saturday, she worked from 9:00 A.M. to 6:00 P.M., with 1 hour off for lunch. Write on Carla's time card the number of hours she worked each day. Then add to find the total number of hours she worked that week.

1.

DAY	A.M.		P.M.		HOURS
	IN	OUT	IN	OUT	
Sunday					
Monday			5:00	9:00	
Tuesday			5:00	9:00	
Wednesday			5:00	9:00	
Thursday			5:00	9:00	
Friday			5:00	9:00	
Saturday	9:00	12:00	1:00	6:00	
				Total Hours	

Carla did not always work the same number of hours each week. Here are Carla's time cards for the next two weeks. Write the number of hours Carla worked each day on her time card. Then add to find the total number of hours she worked each week.

2.

DAY	A.M.		P.M.		HOURS
	IN	OUT	IN	OUT	
Sunday					
Monday			5:00	9:00	
Tuesday			5:00	8:30	
Wednesday			5:00	9:00	
Thursday			5:00	8:30	
Friday			5:00	9:00	
Saturday	9:00	12:00	1:00	5:00	
				Total Hours	

3.

DAY	A.M.		P.M.		HOURS
	IN	OUT	IN	OUT	
Sunday					
Monday			5:00	8:00	
Tuesday			5:00	9:00	
Wednesday			5:00	8:30	
Thursday			5:00	9:00	
Friday			5:00	9:00	
Saturday	9:00	1:00	2:00	5:30	
				Total Hours	

Time Cards

Here are the time cards of some other people who work at Runaway Sports Shop. Figure out how many hours each person worked each day. Write the number of hours on the time card. Then add to find the total number of hours each person worked that week.

1.

DAY	A.M.		P.M.		HOURS
	IN	OUT	IN	OUT	
Sunday			12:00	6:00	
Monday	9:00	1:00	2:00	5:00	
Tuesday	9:00	1:00	2:00	4:30	
Wednesday	9:30	12:30	1:30	5:00	
Thursday	9:00	12:30	1:30	5:30	
Friday	9:00	1:00	2:00	5:30	
Saturday					
				Total Hours	

2.

DAY	A.M.		P.M.		HOURS
	IN	OUT	IN	OUT	
Sunday					
Monday	10:00	12:00	1:00	4:30	
Tuesday	10:00	1:00	2:00	5:00	
Wednesday	10:30	12:00	1:00	5:00	
Thursday	10:00	12:30	1:30	4:30	
Friday	9:00	12:00	1:00	4:00	
Saturday					
				Total Hours	

3.

DAY	A.M.		P.M.		HOURS
	IN	OUT	IN	OUT	
Sunday			12:30	6:30	
Monday	9:30	12:30	1:30	5:00	
Tuesday	9:00	12:00	1:00	5:00	
Wednesday	9:00	12:30	1:30	4:30	
Thursday	9:00	12:00	1:00	5:30	
Friday					
Saturday			1:00	6:00	
				Total Hours	

Addition in Action

Find the answers to these addition problems.

1. 4
    4
    + 4
    ‾‾‾
    12

2. 6
    4
    + 5
    ‾‾‾
    15

3. 8
    9
    + 8
    ‾‾‾

4. 7
    6
    + 6
    ‾‾‾

5. 1
    9
    + 1
    ‾‾‾

6. 7
    3
    3
    + 5
    ‾‾‾

7. 8
    8
    4
    + 9
    ‾‾‾

8. 6
    4
    5
    + 7
    ‾‾‾

9. 8
    2
    2
    + 5
    ‾‾‾

10. 6
    7
    3
    + 6
    ‾‾‾

11. 9
    12
    + 14
    ‾‾‾‾

12. 23
    6
    + 15
    ‾‾‾‾

13. 51
    22
    + 12
    ‾‾‾‾

14. 86
    15
    + 14
    ‾‾‾‾
    115

15. 92
    30
    + 20
    ‾‾‾‾

16. 77
    9
    18
    + 18
    ‾‾‾‾

17. 95
    21
    5
    + 10
    ‾‾‾‾

18. 65
    37
    20
    + 14
    ‾‾‾‾

19. 41
    59
    16
    + 13
    ‾‾‾‾

20. 17
    83
    24
    + 42
    ‾‾‾‾

21. 124
    14
    19
    + 23
    ‾‾‾‾

22. 112
    20
    215
    + 15
    ‾‾‾‾

23. 222
    48
    67
    + 411
    ‾‾‾‾

24. 534
    215
    46
    + 94
    ‾‾‾‾

25. 783
    70
    109
    + 22
    ‾‾‾‾

26. 224
    315
    79
    + 3
    ‾‾‾‾

27. 118
    420
    130
    + 6
    ‾‾‾‾

28. 93
    444
    127
    + 27
    ‾‾‾‾

29. 130
    507
    16
    + 4
    ‾‾‾‾

30. 642
    199
    7
    + 38
    ‾‾‾‾

NAME _____

Earning Money

Each week, Carla added up the hours on her time card. She used the total number of hours to figure out how much money she earned that week.

First she wrote down how much money she earned an hour. Mr. Swift paid her $4.50 an hour. For every hour she worked, she earned:

Then she multiplied that amount by the total number of hours she worked. Here is how she figured her earnings for one week.

$4.50 **amount Carla was paid an hour**
× 27 **number of hours Carla worked**
3150
  900
$121.50 **amount of money Carla earned**

Use Carla's time cards to figure out how much she earned each week. First figure out the total number of hours she worked. Then multiply the number of hours by the amount she earned each hour—$4.50. (Use the Multiplication Table on page 127 if you need help.) Be sure to put the decimal point in the right place in your answers.

1.

DAY	A.M.		P.M.		HOURS
	IN	OUT	IN	OUT	
Sunday					
Monday			5:00	9:00	
Tuesday			5:00	8:00	
Wednesday			5:00	8:30	
Thursday			5:00	9:00	
Friday			5:00	8:30	
Saturday	10:00	12:30	1:30	6:00	
				Total Hours	

Amount of money Carla earned: _____

2.

DAY	A.M.		P.M.		HOURS
	IN	OUT	IN	OUT	
Sunday					
Monday			4:30	9:00	
Tuesday			5:00	9:00	
Wednesday			5:00	8:30	
Thursday			5:00	9:00	
Friday			4:30	8:30	
Saturday	9:00	12:00	1:00	6:00	
				Total Hours	

Amount of money Carla earned:_____

3.

DAY	A.M.		P.M.		HOURS
	IN	OUT	IN	OUT	
Sunday					
Monday			5:00	9:00	
Tuesday			4:30	9:00	
Wednesday			5:00	8:30	
Thursday			5:00	8:30	
Friday			5:00	9:00	
Saturday	9:30	12:00	1:00	6:00	
				Total Hours	

Amount of money Carla earned:_____

4.

DAY	A.M.		P.M.		HOURS
	IN	OUT	IN	OUT	
Sunday					
Monday			5:00	9:00	
Tuesday			6:00	9:00	
Wednesday			5:30	9:00	
Thursday			5:00	8:30	
Friday			4:30	9:00	
Saturday	9:00	12:30	1:00	5:00	
				Total Hours	

Amount of money Carla earned:_____

Figuring Earnings

Here are the time cards of some other people who work at Runaway Sports Shop. The employees do not earn the same amount of money an hour.

Help them to figure out how much they earned each week. First figure out the total number of hours they worked. Then multiply the number of hours by the amount each employee earned an hour.

1. Mary Rossi earns $5.25 an hour.

DAY	A.M.		P.M.		HOURS
	IN	OUT	IN	OUT	
Sunday			12:00	6:00	
Monday	9:30	12:00	1:00	5:00	
Tuesday	9:30	12:30	1:30	5:00	
Wednesday					
Thursday	9:00	12:00	1:00	5:00	
Friday	9:30	12:00	1:00	5:30	
Saturday	9:00	12:00			
				Total Hours	

Amount of money Mary earned: _____

2. Ted Jenkins earns $4.95 an hour.

DAY	A.M.		P.M.		HOURS
	IN	OUT	IN	OUT	
Sunday			1:00	6:00	
Monday			4:30	9:00	
Tuesday			5:00	9:30	
Wednesday			5:00	9:00	
Thursday					
Friday					
Saturday	9:00	12:00	1:00	6:00	
				Total Hours	

Amount of money Ted earned: _____

3. Randy Sumo earns $5.50 an hour.

DAY	A.M.		P.M.		HOURS
	IN	OUT	IN	OUT	
Sunday					
Monday	9:00	12:00	12:30	5:30	
Tuesday	9:00	12:00	12:30	6:00	
Wednesday	9:30	12:00	1:00	6:30	
Thursday			12:30	5:30	
Friday	9:00	12:30	1:00	6:00	
Saturday					
				Total Hours	

Amount of money Randy earned: _____

4. David Weinberg earns $5.75 an hour.

DAY	A.M.		P.M.		HOURS
	IN	OUT	IN	OUT	
Sunday					
Monday			5:00	9:00	
Tuesday			5:30	9:00	
Wednesday	10:00	12:30	1:30	5:00	
Thursday	10:00	12:30	1:00	5:00	
Friday			5:00	9:00	
Saturday					
				Total Hours	

Amount of money David earned: _____

5. Sandy Jackson earns $6.00 an hour.

DAY	A.M.		P.M.		HOURS
	IN	OUT	IN	OUT	
Sunday					
Monday					
Tuesday	9:00	12:00	1:00	5:30	
Wednesday	9:00	12:00	12:30	5:00	
Thursday	9:30	12:30	1:30	6:30	
Friday	9:00	1:00	2:00	6:00	
Saturday	10:00	12:00	1:00	6:00	
				Total Hours	

Amount of money Sandy earned: _____

Figuring Earnings

Every week, Mr. Swift had to figure how much each of his employees had earned. First he wrote down how much the employee was paid an hour. Then he multiplied that amount by the number of hours the employee worked.

Help Mr. Swift figure how much each employee earned a week. Be sure to put the decimal point in the right place in your answers.

1. $4.75
   × 29

   4275
   950

   $137.75

2. $4.25
   × 35

3. $4.95
   × 30

4. $5.65
   × 20

5. $5.90
   × 35

6. $4.25
   × 32

7. $5.10
   × 40

8. $4.50
   × 38

9. $4.35
   × 25

10. $5.85
    × 27

11. $6.00
    × 40

12. $4.35
    × 35

13. $4.30
    × 23

14. $6.25
    × 38

15. $5.50
    × 35

16. $4.45
    × 37

17. $5.65
    × 34

18. $6.10
    × 40

19. $5.25
    × 31

20. $4.85
    × 25

Multiplication in Action

Find the answers to these multiplication problems.
Be sure to put the decimal point in the right place.

1. $2.54
   × 8
   $20.32

2. $9.23
   × 6

3. $8.65
   × 12

4. $1.39
   × 43

5. $3.57
   × 29

6. $9.98
   × 37

7. $7.21
   × 54

8. $4.53
   × 23

9. $6.82
   × 35

10. $1.86
    × 67

11. $4.08
    × 71

12. $5.73
    × 26

13. $8.41
    × 82

14. $2.56
    × 13

15. $7.19
    × 96

16. $8.35
    × 24

17. $1.49
    × 32

18. $7.65
    × 98
    6120
    6885
    $749.70

19. $6.15
    × 75

20. $4.64
    × 59

21. $3.50
    × 62

22. $1.78
    × 89

23. $5.05
    × 48

24. $4.19
    × 16

25. $2.27
    × 41

26. $3.82
    × 83

27. $9.78
    × 56

28. $1.03
    × 87

29. $7.42
    × 64

30. $2.88
    × 21

The Sandwich House

Steve Pollard, one of Carla Williams's friends, also had a part-time job. He worked as a waiter at The Sandwich House. He was paid $4.70 an hour. One day he worked 4 hours. Here's how Steve figured his wages, or earnings, for that day.

$4.70   **hourly wage**
$\underline{\times\ 4}$ **hours worked**
$18.80   **wages earned**

Sometimes the number of hours Steve worked did not come out as a whole number. One day he worked 4½ hours. To figure his wages for 4½ hours, Steve first multiplied his hourly wage by the number of full hours he worked.

$4.70   **hourly wage**
$\underline{\times\ 4}$ **full hours worked**
$18.80   **wages for 4 hours**

To figure his wages for ½ hour, he divided his hourly wage by 2.

$$\begin{array}{r} \underline{\$2.35} \quad \textbf{wages for ½ hour} \\ 2\ \overline{)\ \$4.70} \\ \underline{4} \\ 0\,7 \\ \underline{6} \\ 10 \\ \underline{10} \end{array}$$

Then he added the two amounts together to find his total wages for the day.

$18.80   **wages for 4 hours**
$\underline{+\ 2.35}$ **wages for ½ hour**
$21.15   **wages for 4½ hours**

Here are the hours Steve worked one week. How much did he earn in wages for each day? How much did he earn in wages for the week?

Day	Hours Worked	Wages Earned
Monday	4	$18.80
Tuesday	4½	_____
Wednesday	3½	_____
Thursday	2½	_____
Friday	3	_____
Saturday	7½	_____
	Total wages for the week	_____

Tips

Steve earned $4.70 an hour in wages. But he made more than that because he also got tips from the people he waited on. One day he worked 2 hours and made $19.15. How much did he get in tips that day?

You can use multiplication and subtraction to find out. First, figure how much Steve earned at $4.70 an hour.

$4.70	**hourly wage**
× 2	**hours worked**
$9.40	**wages earned**

Then, to find out how much Steve got in tips, subtract his wages from the total amount.

$19.15	**total**
-$ 9.40	**wages earned**
$ 9.75	**tips**

Here are Steve's hours and earnings for one week. Figure how much he made in tips each day. Then figure how much he made in tips for the whole week.

1.

Day	Hours Worked	Total Earned (wages and tips)	Tips
Monday	2	$19.15	$ 9.75
Tuesday	3	$24.90	_____
Wednesday	3	$26.55	_____
Thursday	2	$17.22	_____
Friday	4	$36.30	_____
Saturday	7	$60.95	_____
		Total tips	_____

Figure Steve's tips for the next week.

1.

Day	Hours Worked	Total Earned (wages and tips)	Tips
Monday	3	$30.00	_____
Tuesday	3	$23.65	_____
Wednesday	2	$15.45	_____
Thursday	2	$19.50	_____
Friday	4	$34.80	_____
Saturday	5	$42.15	_____
		Total tips	_____

Deductions

Carla Williams worked 28 hours her first week at Runaway Sports Shop. She earned $4.50 an hour. So she figured she had earned $126.00.

$$
\begin{array}{rl}
\$4.50 & \textbf{hourly wage} \\
\times\ 28 & \textbf{hours worked} \\
\hline
3600 & \\
900 & \\
\hline
\$126.00 & \textbf{earnings}
\end{array}
$$

But when Carla got her paycheck, she saw that it was made out for only $104.93. Carla asked Mr. Swift why she did not get all the money she had earned. Mr. Swift told Carla that four *deductions* had been taken from her paycheck. The amount of money of each deduction was listed on her paycheck.

DEDUCTIONS

GROSS WAGES	FEDERAL INCOME TAX	STATE INCOME TAX	FICA	SDI	NET WAGES
$126.00	$10.00	$ .10	$9.46	$1.51	$104.93

Mr. Swift explained each thing listed on the paycheck.

Gross wages — This is the amount you earn before any deductions.

Federal income tax — This deduction is used to help run the federal government.

State income tax — This deduction is used to help run your state government.

FICA — (Federal Insurance Contributions Act) The federal government uses the money from FICA deductions for your Social Security payments. If you stop working after age 62, you are sent a Social Security check every month.

SDI — (State Disability Insurance) When you can't work because you are sick or hurt, the money from this deduction will help you until you are able to work again.

Net wages — This is the amount you take home after all deductions have been subtracted from your earnings. This amount is also called *take-home pay*.

Take-Home Pay

Find the net wages, or take-home pay, for each of the
gross wages here. To figure the net wages, add all the
deductions and subtract the total from the gross wages.

	Gross Wages	Federal Income Tax	State Income Tax	FICA	SDI	Net Wages
1.	$ 87.50	$ 4.00	0	$ 6.57	$1.05	$ 75.88
2.	$ 94.50	$ 5.00	0	$ 7.10	$1.13	_____
3.	$ 91.00	$ 5.00	0	$ 6.83	$1.09	_____
4.	$140.40	$13.00	$ .50	$10.54	$1.68	_____
5.	$ 98.80	$ 6.00	0	$ 7.42	$1.19	_____
6.	$145.50	$13.00	$ .50	$10.93	$1.75	_____
7.	$ 96.00	$ 6.00	0	$ 7.21	$1.15	_____
8.	$161.50	$16.00	$ .90	$12.13	$1.94	_____
9.	$136.75	$12.00	$ .30	$10.27	$1.64	_____
10.	$127.25	$10.00	$ .10	$ 9.56	$1.53	_____
11.	$119.90	$ 9.00	0	$ 9.00	$1.44	_____
12.	$131.10	$11.00	$ .30	$ 9.85	$1.57	_____
13.	$123.65	$10.00	$ .10	$ 9.29	$1.48	_____
14.	$150.00	$13.00	$ .50	$11.27	$1.80	_____
15.	$205.25	$22.00	$1.70	$15.41	$2.46	_____
16.	$165.80	$16.00	$ .90	$12.45	$1.99	_____
17.	$187.15	$19.00	$1.30	$14.05	$2.25	_____
18.	$195.00	$20.00	$1.50	$14.64	$2.34	_____
19.	$210.50	$24.00	$2.10	$15.81	$2.53	_____
20.	$227.35	$25.00	$2.50	$17.07	$2.73	_____

What's the Bill?

Carla enjoyed her job at Runaway Sports Shop. She liked helping people and selling sporting goods. Below are some of the things Carla sold. Help her figure the total cost of the orders below. Write the total in the blank next to each order.

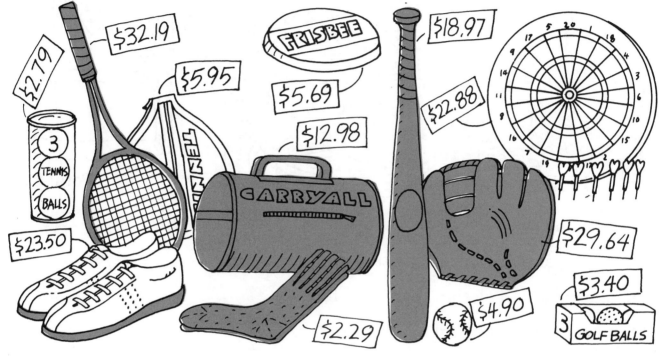

Total Cost

1. 2 baseballs, 1 baseball bat, 1 carryall bag $ 41.75

2. 1 dart game, 3 Frisbees, 2 packs of golf balls   _____

3. 2 tennis rackets, 2 racket covers, 2 cans of tennis balls   _____

4. 1 baseball glove, 1 pair of shoes, 3 baseballs   _____

5. 3 racket covers, 1 carryall bag, 3 pairs of socks   _____

6. 2 dart games, 1 Frisbee, 4 cans of tennis balls   _____

7. 2 baseball gloves, 4 baseballs, 1 pair of socks   _____

8. 4 packs of golf balls, 2 carryall bags, 4 pairs of socks   _____

9. 3 pairs of shoes, 6 pairs of socks, 1 baseball   _____

10. 2 Frisbees, 1 dart game, 3 cans of tennis balls   _____

	Total Cost
11. 1 fishing pole, 1 net, 3 dozen fish hooks	———
12. 2 backpacks, 1 canteen, 2 sets of dishes	———
13. 2 compasses, 1 sleeping bag, 1 flashlight	———
14. 1 backpack, 2 flashlights, 4 batteries	———
15. 3 sets of dishes, 1 compass, 2 dozen fish hooks	———
16. 2 canteens, 2 sleeping bags, 3 batteries	———
17. 4 flashlights, 8 batteries, 2 nets	———
18. 5 backpacks, 5 canteens	———
19. 3 compasses, 3 flashlights	———
20. 3 fishing poles, 4 dozen fish hooks	———
21. 6 sets of dishes, 6 canteens	———
22. 1 sleeping bag, 2 flashlights, 2 batteries	———
23. 3 nets, 3 fishing poles, 5 dozen fish hooks	———
24. 4 backpacks, 4 sleeping bags	———
25. 2 fishing poles, 2 canteens, 2 compasses	———

Making Change

As a waiter, Steve had to know how to make change correctly. Here are some of his orders. Figure how much change Steve should give back for each order. Write the correct change in the blanks.

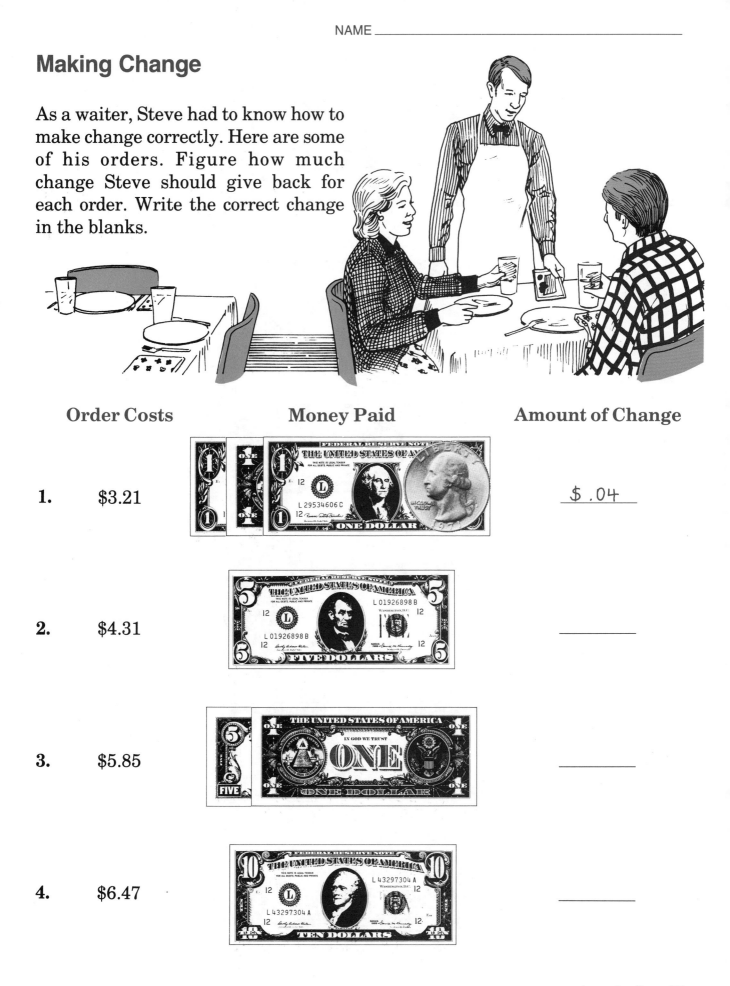

Order Costs	Money Paid	Amount of Change
1. $3.21		$ .04
2. $4.31		_____
3. $5.85		_____
4. $6.47		_____

Order Costs	Money Paid	Amount of Change
5. $ 3.97	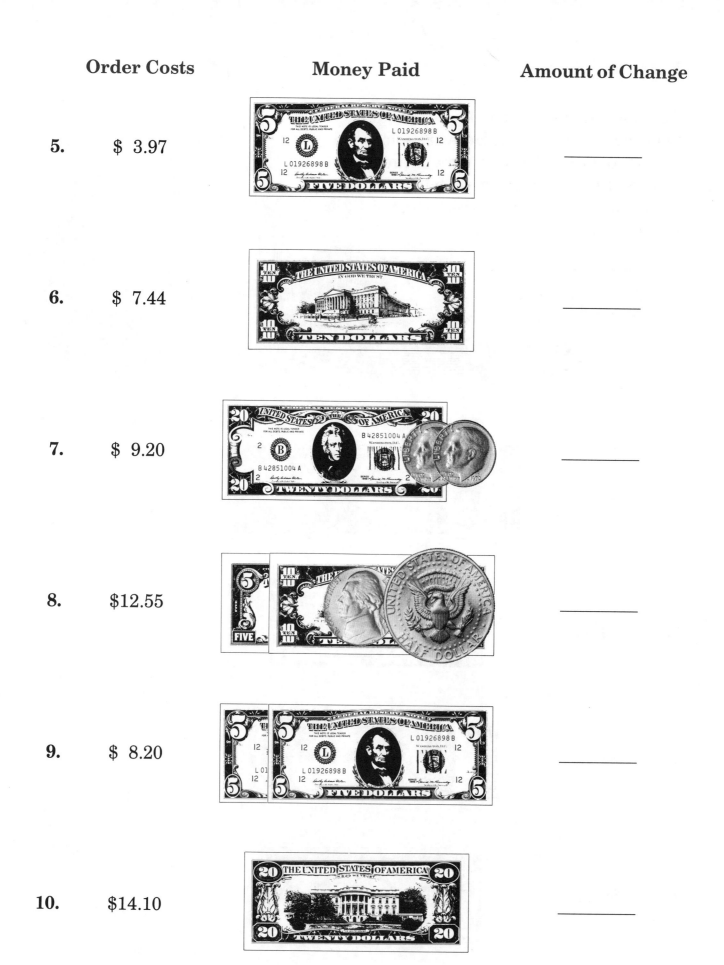	_____
6. $ 7.44		_____
7. $ 9.20		_____
8. $12.55		_____
9. $ 8.20		_____
10. $14.10		_____

How Much Is There?

Mr. Swift asked Carla to find out how many tennis balls the store had altogether. Carla knew that there were 3 balls in each can. She knew that there were 12 cans in each case.

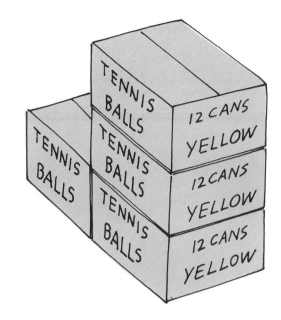

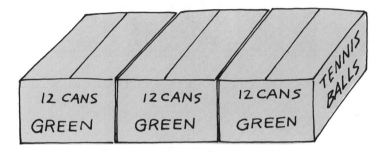

Use the picture to help Carla answer the questions below. Write your answers in the blanks.

1. How many cases of yellow tennis balls are there? ____4____

2. How many cans of yellow tennis balls are there altogether? _____

3. How many yellow tennis balls are there altogether? _____

4. How many cases of green tennis balls are there? _____

5. How many cans of green tennis balls are there altogether? _____

6. How many green tennis balls are there altogether? _____

7. Each can of tennis balls sells for $2.79. What is the price of each ball? _____

8. What is the selling price of each case? _____

9. What is the total selling price of *all* the cases? _____

10. To make a profit, Mr. Swift sells the tennis balls for more than he paid for them. He paid $187.49 for all the cases. How much more is his selling price for all the cases? _____

Figuring the Cost

Runaway Sports Shop was selling tickets to a special baseball game. Figure the total cost of each group of tickets below. Write the total in the blank.

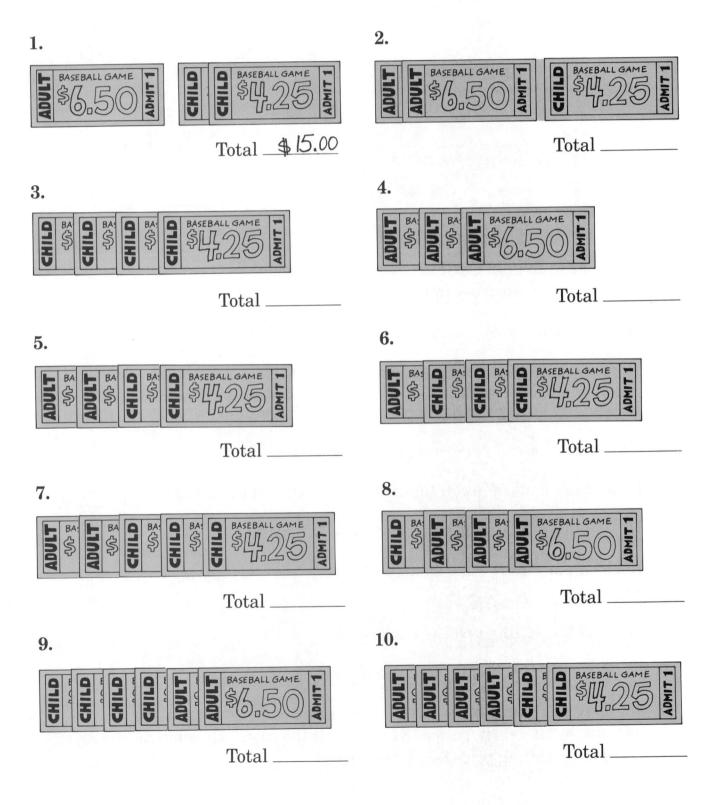

1.

Total ___$15.00___

2.

Total _____

3.

Total _____

4.

Total _____

5.

Total _____

6.

Total _____

7.

Total _____

8.

Total _____

9.

Total _____

10.

Total _____

Sports Store Maze

Start at the top left square of the maze. Follow the
openings through the squares. As you go, add the
prices of the things Carla sold. (Do your work on another
piece of paper.) Then fill in the blank for the total amount
at the end of the maze.

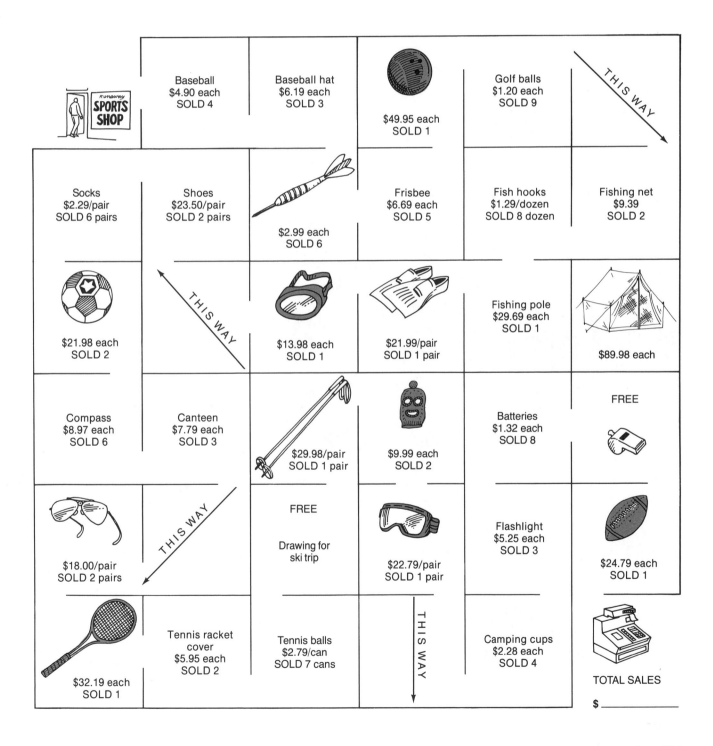

TOTAL SALES

$ _____

The Wage Wheel

All employees of Runaway Sports Shop earned the same hourly wage when they first started their jobs. Figure the daily wages, weekly wages, and yearly wages for the six employees below. First, count the money in the center of the wheel. Then work outward toward the rim, filling in the blanks as you go. (As you work toward the rim, each answer is part of the next problem.)

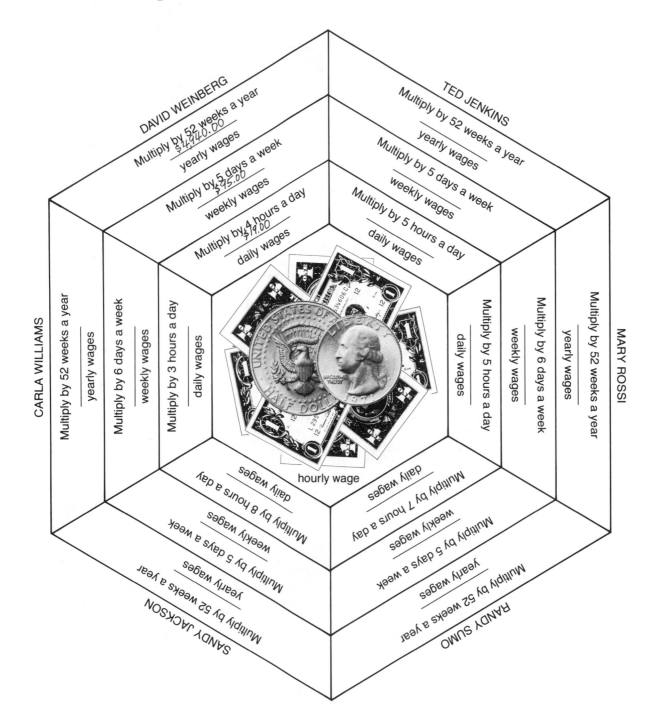

DAVID WEINBERG
Multiply by 52 weeks a year — $4,940.00 yearly wages
Multiply by 5 days a week — $95.00 weekly wages
Multiply by 4 hours a day — $19.00 daily wages

TED JENKINS
Multiply by 52 weeks a year — yearly wages
Multiply by 5 days a week — weekly wages
Multiply by 5 hours a day — daily wages

CARLA WILLIAMS
Multiply by 52 weeks a year — yearly wages
Multiply by 6 days a week — weekly wages
Multiply by 3 hours a day — daily wages

MARY ROSSI
Multiply by 52 weeks a year — yearly wages
Multiply by 6 days a week — weekly wages
Multiply by 5 hours a day — daily wages

hourly wage

SANDY JACKSON
Multiply by 52 weeks a year — yearly wages
Multiply by 5 days a week — weekly wages
Multiply by 8 hours a day — daily wages

RANDY SUMO
Multiply by 52 weeks a year — yearly wages
Multiply by 5 days a week — weekly wages
Multiply by 7 hours a day — daily wages

Half-Price Sale!

The Sandwich House was having a grand opening sale.
For one day only, everything would be sold for half
the regular price. Below are some of Steve's orders. To
find the half price of each order, first figure the total,
using the regular price. Then divide this amount by 2. Be
sure to put the decimal point in the right place. Write
your answers in the blanks.

The Sandwich House

Ham sandwich	$3.30	Soup	$1.75
Turkey sandwich	3.20	Salad	2.20
Fish sandwich	3.05	Milk, soft drinks	.85
Cheese sandwich	2.75	Coffee	.75
Chicken sandwich	3.15	Pie	1.90
Fruit plate	3.50	Ice cream	1.35

	Order	Total	Half Price
1.	1 ham sandwich, 1 soup, 1 milk	$5.90	$2.95
2.	1 pie, 2 coffees	_____	_____
3.	1 cheese sandwich, 1 soft drink	_____	_____
4.	1 fruit plate, 1 milk, 1 ice cream	_____	_____
5.	1 chicken sandwich, 1 coffee	_____	_____
6.	1 soup, 1 soft drink	_____	_____
7.	1 fish sandwich, 1 milk, 1 pie	_____	_____

Order		Total	Half Price
8.	2 turkey sandwiches, 1 salad, 2 soft drinks	_____	_____
9.	1 fish sandwich, 1 chicken sandwich, 2 coffees	_____	_____
10.	1 cheese sandwich, 1 milk, 1 pie	_____	_____
11.	1 fruit plate, 1 soft drink, 1 ice cream	_____	_____
12.	2 ham sandwiches, 1 salad, 2 soft drinks	_____	_____
13.	2 chicken sandwiches, 2 soups, 2 milks	_____	_____
14.	3 fish sandwiches, 3 coffees	_____	_____
15.	1 ham sandwich, 1 chicken sandwich, 1 soup, 2 milks	_____	_____
16.	2 cheese sandwiches, 2 salads, 2 soft drinks	_____	_____
17.	4 turkey sandwiches, 2 milks, 2 coffees	_____	_____
18.	3 chicken sandwiches, 2 soups, 3 soft drinks	_____	_____
19.	2 fruit plates, 1 soup, 3 milks	_____	_____
20.	1 cheese sandwich, 1 ham sandwich, 3 coffees	_____	_____
21.	3 soups, 2 salads, 3 soft drinks	_____	_____
22.	5 ice creams, 5 coffees	_____	_____
23.	1 cheese sandwich, 1 fruit plate, 1 soup, 2 milks	_____	_____
24.	3 cheese sandwiches, 2 soups, 3 soft drinks	_____	_____
25.	4 ham sandwiches, 4 salads, 4 milks	_____	_____
26.	2 fruit plates, 3 soups, 3 soft drinks	_____	_____
27.	5 chicken sandwiches, 5 coffees, 2 ice creams	_____	_____
28.	3 fruit plates, 3 turkey sandwiches, 6 milks	_____	_____
29.	3 cheese sandwiches, 3 soups, 4 soft drinks	_____	_____
30.	4 pies, 3 ice creams, 7 coffees	_____	_____

Shopping Quiz

Both Steve and Carla needed to buy clothes for their new jobs. They saw this ad for Gant's Department Store in the newspaper. Read the ad. Then answer the questions on the next page.

GANT'S SUPER SALE
BUY ANY QUANTITY AT THESE SPECIAL SALE PRICES

Men's Socks
5 pairs/$9.95
(REGULAR PRICE — $11)

Any 2 Belts for $16.98
(REGULAR PRICE—$9 EACH)

Shirts
3 for $39.00
(REG.—$16 EACH)

Women's Socks
4 pairs/$10
(REGULAR PRICE—$3.50 A PAIR)

Save $12 When You Buy 2!
Sweaters 2 for $44.00

Men's and Women's Shoes
Any 2 pairs for $65
(REG. $35 A PAIR)

SAVE $2 ON EVERY TIE!
Men's Ties
2/$15

Women's Scarves
3/$19.95
(REG.—$7 EACH)

1. What is the sale price of one pair of men's socks? $ 1.99

   How much do four pairs cost? $ 7.96

2. What is the sale price of one belt?   _____

   How much do three belts cost?   _____

3. What is the sale price of one shirt?   _____

   How much do two shirts cost?   _____

4. How much do four pairs of women's socks cost at the regular price?   _____

   How much money is saved by buying four pairs at the sale price?   _____

5. What is the sale price of one sweater?   _____

   How much more money does one sweater cost at the regular price?   _____

6. What is the sale price of one pair of shoes?   _____

   How much money is saved by buying two pairs of shoes at the sale price?   _____

7. What is the sale price of one tie?   _____

   What is the regular price of one tie?   _____

8. What is the sale price of one scarf?   _____

   How much do four scarves cost?   _____

9. At the sale, Steve bought three pairs of men's socks, two shirts, and three ties. How much money did he spend?   _____

   How much money did he save by buying the clothes on sale?   _____

   If Steve pays with three twenty-dollar bills, how much change will he get?   _____

10. At the sale, Carla bought two sweaters, one pair of shoes, and two scarves. How much money did she spend?   _____

    How much money did she save by buying these things on sale?   _____

    If Carla pays with six twenty-dollar bills, how much change will she get?   _____

Division in Action

Find the answers to these division problems. Be sure to put the decimal point in the right place.

1.
$$
\begin{array}{r}
\$.93 \\
2\overline{)\$1.86} \\
\underline{18} \\
6 \\
\underline{6}
\end{array}
$$

2. $2\overline{)\$32.64}$

3. $4\overline{)\$39.00}$

4. $3\overline{)\$5.67}$

5. $7\overline{)\$14.07}$

6. $4\overline{)\$27.76}$

7.
$$
\begin{array}{r}
\$3.58 \\
3\overline{)\$10.74} \\
\underline{9} \\
17 \\
\underline{15} \\
24 \\
\underline{24}
\end{array}
$$

8. $5\overline{)\$43.25}$

9. $8\overline{)\$68.56}$

10. $6\overline{)\$7.32}$

11. $9\overline{)\$36.36}$

12. $2\overline{)\$7.28}$

13. $4\overline{)\$14.92}$

14. $6\overline{)\$44.76}$

15. $5\overline{)\$36.75}$

16. $9\overline{)\$24.93}$

17. $3\overline{)\$13.02}$

18. $5\overline{)\$11.40}$

19. $7\overline{)\$23.94}$

20. $4\overline{)\$7.16}$

21. $2\overline{)\$7.70}$

22. $6\overline{)\$58.92}$

23. $8\overline{)\$79.68}$

24. $5\overline{)\$97.15}$

The Wage Wheel

Figure the weekly wages, daily wages, and hourly wages for the six employees of Gant's below. Start with the monthly wages and work outward toward the rim, filling in the blanks as you go. (As you work toward the rim, each answer is part of the next problem.)

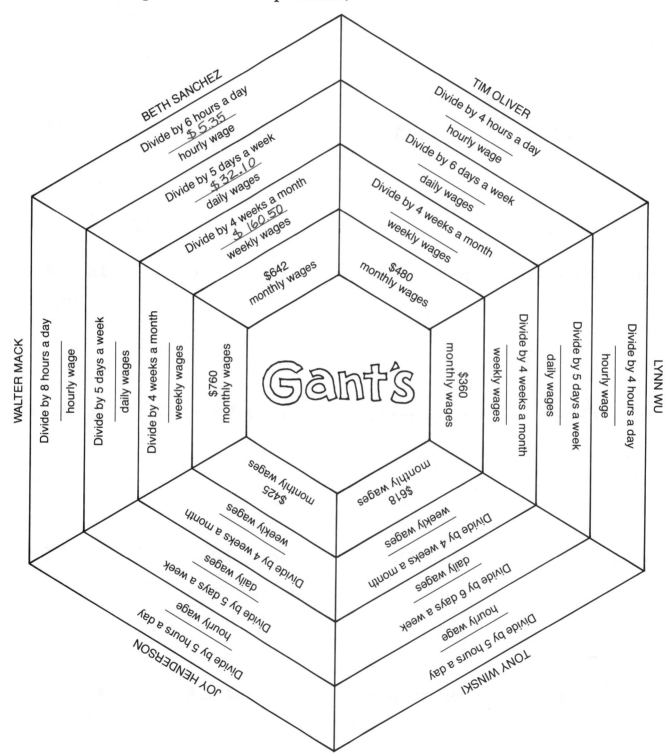

BETH SANCHEZ
- Divide by 6 hours a day — $5.35 — hourly wage
- Divide by 5 days a week — $32.10 — daily wages
- Divide by 4 weeks a month — $160.50 — weekly wages
- $642 monthly wages

TIM OLIVER
- Divide by 4 hours a day — hourly wage
- Divide by 6 days a week — daily wages
- Divide by 4 weeks a month — weekly wages
- $480 monthly wages

WALTER MACK
- Divide by 8 hours a day — hourly wage
- Divide by 5 days a week — daily wages
- Divide by 4 weeks a month — weekly wages
- $760 monthly wages

LYNN WU
- Divide by 4 hours a day — hourly wage
- Divide by 5 days a week — daily wages
- Divide by 4 weeks a month — weekly wages
- $360 monthly wages

JOY HENDERSON
- Divide by 5 hours a day — hourly wage
- Divide by 5 days a week — daily wages
- Divide by 4 weeks a month — weekly wages
- $425 monthly wages

TONY WINSKI
- Divide by 5 hours a day — hourly wage
- Divide by 6 days a week — daily wages
- Divide by 4 weeks a month — weekly wages
- $618 monthly wages

Figuring Mileage

Bob Sommers had a full-time job at Franklin Florist. His job was to deliver flowers. One morning the owner, Ms. Franklin, asked Bob to find out how many miles the company station wagon got per gallon. He needed to figure out how many miles the station wagon could go on one gallon of gas.

"I have a full tank of gas now," Bob said. "I'll keep track of how many miles I drive today. Then I'll buy more gas and see how many gallons I used. By dividing the number of miles driven by the number of gallons used, I'll know how many miles I get per gallon."

Before Bob started driving, he wrote down the mileage reading on the car's odometer.

This was the mileage on the odometer at the start of the day.

`3 7 8 2 5`

This was the mileage on the odometer at the end of the day.

`3 7 8 9 4`

To find out how many miles he had gone, Bob subtracted his first mileage reading from the last one.

$$
\begin{array}{r}
37894 \\
- 37825 \\
\hline
69
\end{array}
$$
mileage at end of trip
mileage at start of trip
miles driven

Then Bob filled the tank again. It took 3 gallons. To find out how many miles he got per gallon, Bob divided the number of miles driven by the number of gallons of gas used.

$$
\begin{array}{r}
23 \\
3 \overline{) 69}
\end{array}
$$
miles per gallon
gallons   **miles driven**

Now Bob knew how many miles the station wagon got per gallon. He knew he could drive about 23 miles on one gallon of gas.

Figure the gas mileage for the cars here. Write the number of miles each car was driven before filling up again. Then find the number of miles per gallon each car gets.

	Odometer at Start	Odometer at End	Miles Driven	Gallons Needed to Fill Tank Again	Miles per Gallon
1.	41435	41580	145	5	29
2.	48763	48889		6	
3.	58354	58515		7	
4.	16249	16409		10	
5.	18716	18796		4	
6.	32290	32395		3	
7.	54178	54303		5	
8.	42752	42860		6	
9.	57279	57423		6	
10.	27004	27300		8	
11.	34148	34276		4	
12.	23511	23731		10	
13.	42632	42762		5	
14.	71807	71959		4	
15.	44427	44598		9	
16.	26692	26794		6	
17.	17886	18110		8	
18.	40635	40873		7	
19.	33833	33914		3	
20.	12909	13119		10	

Decimal Division

One week, all Bob's driving was done on the highway. He knew that he probably got better gas mileage on the highway than in stop-and-go city driving. So he decided to find out for sure.

He checked the odometer when the tank was full. After a week of highway driving, he checked it again. By subtracting the first odometer reading from the second, Bob knew the car had been driven 369 miles.

First fill-up **3 8 2 7 6**

Second fill-up **3 8 6 4 5**

When the tank was filled again, Bob saw how many gallons he needed.

GALLONS: $14\frac{6}{10}$

To find the miles per gallon, Bob had to divide 369 by 14-6/10.

gallons $14\frac{6}{10}\,\overline{)\,369}$ miles driven

First he changed the 6/10 to a decimal fraction. He knew that 6/10 and .6 are just different ways of writing the same fraction.

six-tenths $= \frac{6}{10} = .6$

After he changed 6/10 to a decimal fraction, the problem looked like this:

$14.6\,\overline{)\,369}$

Next, Bob moved the decimal point one place to the right in both numbers. He added two more zeros to the 369. Adding zeros to the right of the decimal point doesn't change the value of the number. Now he was ready to divide.

$14\underset{\smile}{6}.\,\overline{)\,369\underset{\smile}{0}.00}$

Bob worked out the problem to two decimal places. This means he stopped dividing when his answer had two numbers after the decimal point. He found that the station wagon got about 25.27 miles per gallon on the highway. This was over 2 miles per gallon more than in city driving.

$$
\begin{array}{r}
2\,5.27 \text{ miles per gallon} \\
14\underset{\smile}{6}.\,\overline{)\,369\underset{\smile}{0}.00} \\
\underline{292} \\
77\,0 \\
\underline{73\,0} \\
4\,0\,0 \\
\underline{2\,9\,2} \\
1\,0\,80 \\
\underline{1\,0\,22} \\
58
\end{array}
$$

Change these fractions to decimal fractions.

1. $12\frac{3}{10}$ _12.3_ 2. $11\frac{7}{10}$ _____ 3. $9\frac{6}{10}$ _____ 4. $14\frac{2}{10}$ _____ 5. $15\frac{9}{10}$ _____

6. $13\frac{4}{10}$ _____ 7. $8\frac{8}{10}$ _____ 8. $6\frac{1}{10}$ _____ 9. $10\frac{5}{10}$ _____ 10. $7\frac{7}{10}$ _____

11. $12\frac{8}{10}$ _____ 12. $9\frac{3}{10}$ _____ 13. $14\frac{7}{10}$ _____ 14. $13\frac{8}{10}$ _____ 15. $15\frac{2}{10}$ _____

Find the answers to these division problems. Remember to move the decimal point one place to the right before dividing. Work out each problem to two decimal places.

16.
```
        15.95
8.4 )1340.0
     84
     500
     420
      800
      756
      440
      420
       20
```
17. $9.6\,)\overline{131}$ 18. $8.5\,)\overline{107}$ 19. $8.6\,)\overline{106}$

20. $5.8\,)\overline{120}$ 21. $6.5\,)\overline{32}$ 22. $12.1\,)\overline{156}$ 23. $14.5\,)\overline{238}$

24. $11.7\,)\overline{177}$ 25. $13.2\,)\overline{211}$ 26. $15.4\,)\overline{223}$ 27. $13.4\,)\overline{163}$

28. $10.8\,)\overline{150}$ 29. $8.5\,)\overline{127}$ 30. $13.9\,)\overline{189}$ 31. $5.4\,)\overline{96}$

32. $10.6\,)\overline{230}$ 33. $11.8\,)\overline{172}$ 34. $12.9\,)\overline{224}$ 35. $14.8\,)\overline{230}$

Going Places

One day, Ms. Franklin asked Bob to drive to Columbia International Airport. He needed to pick up fresh flowers from Hawaii.

Bob got out a map and found that the airport was 162 miles away. He knew that the station wagon got about 25.3 miles per gallon of gas on the highway. How many gallons would the station wagon need to get to the airport? To find out, Bob divided the distance to the airport by the number of miles per gallon the car got.

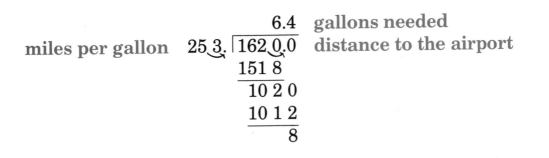

```
                          6.4   gallons needed
miles per gallon    25.3. |162 0.0   distance to the airport
                          151 8
                          10 2 0
                          10 1 2
                              8
```

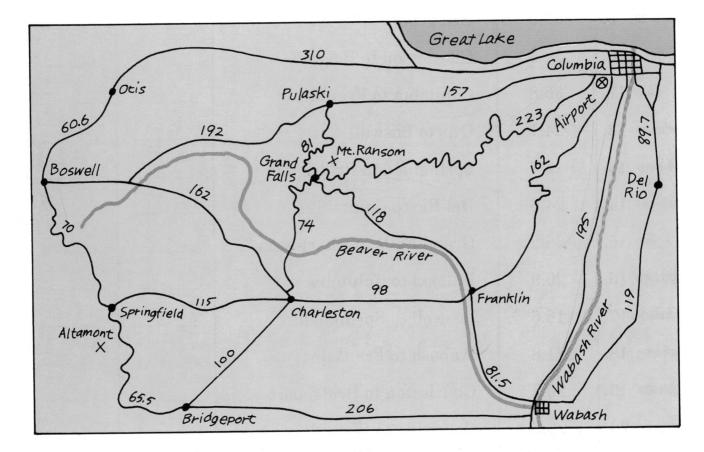

Use the map on page 39 to help you figure out how many
gallons of gas each car would need to make these trips.
Work out each problem to one decimal place.

		Miles per Gallon	Trip	Distance	Gallons Needed
	1.	20.9	Franklin to Grand Falls	118	5.6
	2.	28.2	Grand Falls to Columbia		
	3.	34.8	Columbia to Otis		
	4.	28.3	Wabash to Del Rio		
	5.	17.8	Franklin to Charleston		
	6.	19.3	Boswell to Pulaski		
	7.	36.9	Springfield to Charleston		
	8.	26.1	Grand Falls to Pulaski		
	9.	21.9	Bridgeport to Wabash		
	10.	32.4	Charleston to Boswell		
	11.	36.8	Columbia to Franklin		
	12.	24.1	Otis to Boswell		
	13.	18.7	Springfield to Bridgeport		
	14.	24.6	Del Rio to Columbia		
	15.	35.2	Grand Falls to Charleston		
	16.	20.9	Pulaski to Columbia		
	17.	16.6	Boswell to Springfield		
	18.	21.8	Wabash to Franklin		
	19.	26.3	Charleston to Bridgeport		
	20.	29.5	Columbia to Wabash		

Renting a Car

One day, the Franklin Florist station wagon broke down. Ms. Franklin asked Bob to find out the cost for renting a small station wagon for the day.

Bob called two places that rented cars. He knew that he would drive about 75 miles. He used that figure to compare costs.

Call-a-Car charged $44.95 a day with free mileage. No matter how many miles Bob drove the car, it would still cost him $44.95.

$44.95 total cost (Call-a-Car)

Great Lake Rent-a-Car charged $24.00 a day and $.20 a mile. Bob multiplied $.20 by 75 miles to find the total mileage charge. Then he added this amount to the $24.00 daily charge to get the total cost.

$$\begin{array}{r} \$\ .20 \\ \times\ \ 75 \\ \hline 100 \\ 140 \\ \hline \$15.00 \end{array}$$ cost per mile
miles

total mileage charge

$$\begin{array}{r} \$15.00 \\ +\ 24.00 \\ \hline \$39.00 \end{array}$$ total mileage charge
daily charge
total cost
(Great Lake Rent-a-Car)

Bob compared the total costs of each company. Great Lake Rent-a-Car had the lower total cost. To find out how much the car cost per mile, Bob divided the total cost by the number of miles he planned to drive. It cost about $.52 a mile to rent a car from Great Lake Rent-a-Car.

$$75\overline{\left)\begin{array}{r} \$\ .52 \\ \$39.00 \\ \underline{37\ 5} \\ 1\ 50 \\ \underline{1\ 50} \end{array}\right.}$$ cost per mile

In each problem below, find out if Company A or Company B has the lower cost for renting a car. Use the total miles to be driven to compare costs. When you have found the lower cost for each problem, figure the cost per mile. Round your answer to two decimal places.

	Total Miles	Company A	Company B	Lower Total Cost	Lower Cost per Mile
1.	100	$35.00 a day Free mileage Total cost _$35.00_	$23.00 a day $.22 a mile Total cost _$45.00_	Company ____ $ _35.00_	$ _.35_
2.	75.5	$21.95 a day $.22 a mile Total cost _38.56_	$39.95 a day Free mileage Total cost ____	Company ____ $ ____	$ ____
3.	54.2	$47.00 a day Free mileage Total cost ____	$22.77 a day $.25 a mile Total cost ____	Company ____ $ ____	$ ____
4.	38.6	$25.95 a day Free mileage Total cost ____	$22.50 a day $.25 a mile Total cost ____	Company ____ $ ____	$ ____
5.	71.5	$22.50 a day $.26 a mile Total cost ____	$35.25 a day Free mileage Total cost ____	Company ____ $ ____	$ ____
6.	98.3	$24.95 a day $.20 a mile Total cost ____	$36.50 a day Free mileage Total cost ____	Company ____ $ ____	$ ____
7.	45.5	$24.75 a day $.24 a mile Total cost ____	$42.95 a day Free mileage Total cost ____	Company ____ $ ____	$ ____
8.	30.8	$21.50 a day $.30 a mile Total cost ____	$28.75 a day Free mileage Total cost ____	Company ____ $ ____	$ ____
9.	66.5	$22.00 a day $.26 a mile Total cost ____	$41.95 a day Free mileage Total cost ____	Company ____ $ ____	$ ____
10.	120	$21.25 a day $.23 a mile Total cost ____	$41.00 a day Free mileage Total cost ____	Company ____ $ ____	$ ____

Buying Tires

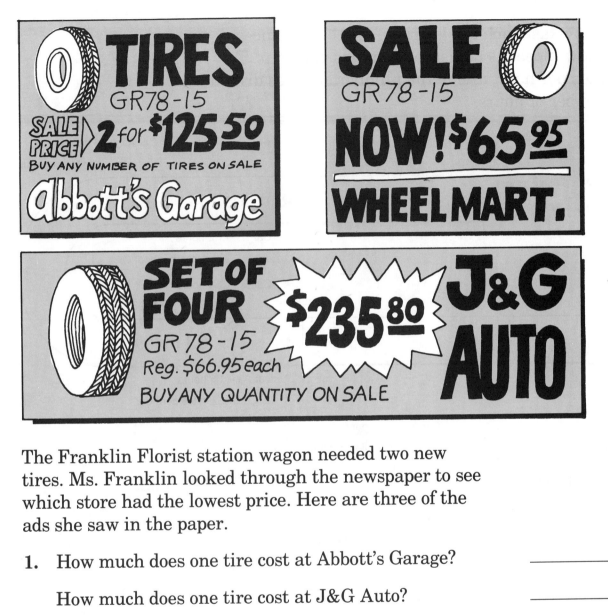

The Franklin Florist station wagon needed two new tires. Ms. Franklin looked through the newspaper to see which store had the lowest price. Here are three of the ads she saw in the paper.

1. How much does one tire cost at Abbott's Garage? _____

    How much does one tire cost at J&G Auto? _____

2. At which store would Ms. Franklin get the lowest

    price if she wanted to buy only one tire? _____

3. How much do two tires cost at Wheelmart? _____

    How much do two tires cost at J&G Auto? _____

4. Which store has the lowest price for two tires? _____

5. If Ms. Franklin buys two tires at J&G Auto, how _____

    much will she save by buying them on sale? _____

Take a Trip

Find out how fast each car is driven for the distances shown below. Speed is measured in miles per hour. To figure miles per hour, divide the number of miles driven by the number of hours.

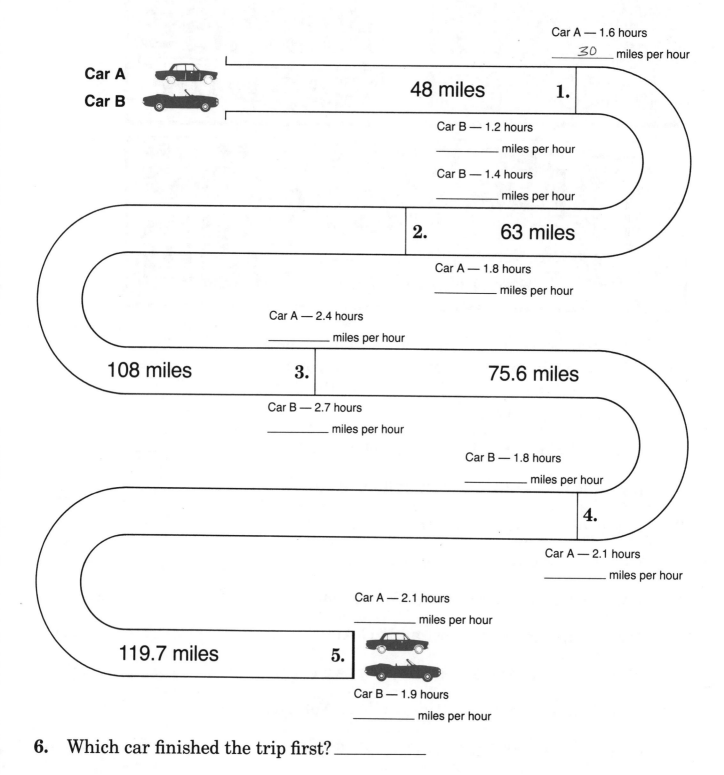

Car A — 1.6 hours

_____30_____ miles per hour

Car A
Car B

48 miles

1.

Car B — 1.2 hours

_____ miles per hour

Car B — 1.4 hours

_____ miles per hour

2. 63 miles

Car A — 1.8 hours

_____ miles per hour

Car A — 2.4 hours

_____ miles per hour

108 miles **3.** 75.6 miles

Car B — 2.7 hours

_____ miles per hour

Car B — 1.8 hours

_____ miles per hour

4.

Car A — 2.1 hours

_____ miles per hour

Car A — 2.1 hours

_____ miles per hour

119.7 miles **5.**

Car B — 1.9 hours

_____ miles per hour

6. Which car finished the trip first? _____

A Checking Account

Soon after Carla Williams started working at Runaway Sports Shop, she opened a checking account at the bank. She knew that having a checking account was a good way to handle her money.

Here are some of the reasons a checking account is good to have.

- You don't have to carry a lot of money with you to pay for things. In many stores, you can write a check to pay for what you buy.
- Checks are a safe and easy way to pay bills by mail. No one can cash a check except the person, store, or company to whom it is made out. Money sent by mail can get lost or stolen.
- When you write a check, there is a record that you paid for what you bought. A copy of the canceled check can be your written record.

Writing a check to pay for something is like paying cash for it. But the money is in your checking account and not in your pocket. You must have enough money in your checking account to cover the check. Your bank will pay the person, store, or company you name on the check with the money in your account.

After your bank pays the money for each check you write, the check is *canceled*. A canceled check cannot be used again. At the end of the month, your bank sends you a *statement* that lists all your canceled checks. If you need a copy of a canceled check, you can contact your bank. The copy will be your record that you paid for the things you bought.

Answer these questions YES or NO.

1. Can Carla write a check to pay for something in a store? _____

2. Are canceled checks a record that you paid for what you bought? _____

3. Can a canceled check be used again? _____

4. Should Carla have enough money in her account to cover the checks she

   writes? _____

Complete each sentence by drawing a line from the left
to the best ending on the right.

                                  carry a lot of money with her.

5. Carla opened a checking account
   so she could

                                  write checks.

                                  send the money.

6. A good way to pay bills
   by mail is to

                                  send a check.

                                  paying cash for it.

7. Writing a check to pay
   for something is like

                                  saving money to pay for it.

                                  paid for the thing you bought.

8. A copy of a canceled check
   shows that you

                                  don't have any money in your account.

Understanding Checks

Carla bought two Frisbees from Runaway Sports Shop.
Here is the check she wrote to pay for them.

① **UNITED NATIONAL BANK**
Hometown, Illinois

NUMBER 101
09-38
241

October 15 19 89

PAY TO
THE ORDER OF ___ Runaway Sports Shop ___ $ 14 45

___ Fourteen and 45/100 ___ DOLLARS
nonnegotiable

Carla Williams
7294 West Street
Hometown, Illinois

Carla Williams

⑩ ② ③ ④ ⑤ ⑥ ⑦ ⑧ ⑨

Here are the things that are on this check.

① The name of the bank where Carla has her checking account.

② The name of the company to whom Carla made the check out. Runaway Sports Shop will cash this check to get its money.

③ The number of the check. The checks are numbered in order, so the next check Carla writes will be Number 102.

④ The bank number. Every bank has its own number above the line. (A government bank number is below the line.)

⑤ The date. Carla wrote this check on October 15.

⑥ The amount of the check written in numbers. Carla wrote this check for $14.45

⑦ The dollar amount of the check spelled out in words.

⑧ Carla's signature. Carla must sign her name on the check before her bank will cash it and pay Runaway Sports Shop.

⑨ Carla's account number.

⑩ Carla's name and address. This shows that the check belongs to Carla.

Here is another check. Study it and answer
the questions below.

1.   Who wrote this check? _____

2.   At what bank does Sylvia Krebs have her checking account? _____

_____

3.   Where does Sylvia Krebs live? _____

_____

4.   To whom did Sylvia Krebs make out this check? _____

_____

5.   What is the number of this check? _____

6.   On what date was this check written? _____

7.   What is the amount of this check? _____

8.   In how many places on the check is the amount shown? _____

9.   What is the bank number? _____

Writing Checks

When Carla opened her checking account, the bank gave her a list of rules for writing checks. These rules showed her the correct way to write checks.

RULES FOR WRITING CHECKS

1. Always use a pen to write a check. Never write a check with a pencil.
2. Never change figures on a check. If you make a mistake, tear up the check and write a new one.
3. Fill all the blanks on the check with words, numbers, or lines.
4. Be sure to use the correct date.
5. Never write a check for more than the amount in your checking account.
6. If a check is lost or stolen, ask the bank to stop payment on it.

Carla also learned how to write the amount on the check both in figures and in words.

Write the amount in figures close to the dollar sign ($) on the check. On the next line, write the amount of dollars in words. Then write *and.* Then write the amount of cents in hundredths.

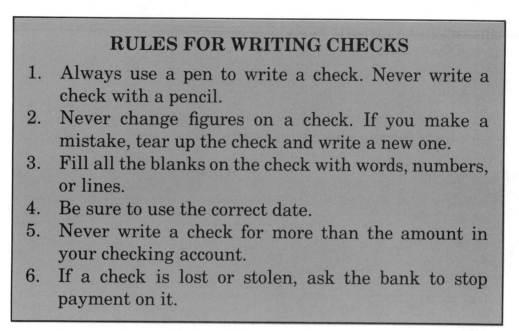

The amount of the next check Carla
wrote was $9.67.

She wrote the amount in figures like this. $ _9 67_

She wrote the amount in words like this.   _Nine and 67/100_ _____ DOLLARS

Write these amounts for the checks in words.

Amount	Words
1. $8.52	_Eight and 52/100_ _____ DOLLARS
2. $115.69	_One Hundred Fifteen and 69/100_ _____ DOLLARS
3. $71.80	_____ DOLLARS
4. $16.97	_____ DOLLARS
5. $10.22	_____ DOLLARS
6. $49.51	_____ DOLLARS
7. $20.82	_____ DOLLARS
8. $339.17	_____ DOLLARS
9. $25.12	_____ DOLLARS
10. $212.39	_____ DOLLARS
11. $31.40	_____ DOLLARS
12. $52.33	_____ DOLLARS
13. $164.10	_____ DOLLARS
14. $19.18	_____ DOLLARS
15. $97.05	_____ DOLLARS

Writing Checks

Carla bought new clothes at Gant's Department Store.
She wrote this check to pay for the clothes.

Study the check that Carla wrote. Then make out the
checks on this page and the next for the amounts shown.
Use today's date and sign your own name.

1. To Jennifer Shikamura for $32.28.

2. To Kaplan's Electronics for $116.59.

Avenue Bank

Newark, Delaware

Number 188

01-16 / 815

19____

PAY TO THE ORDER OF _____ $ _____

_____ DOLLARS

nonnegotiable

⑈0815⑈ 0116⑈ 49

3. To Entropy Power Company for $19.07.

NATIONAL BANK OF TAVARES

TAVARES, FLORIDA

Number 355

06-38 / 105

_____ 19____

PAY TO THE ORDER OF _____ $ _____

_____ DOLLARS

nonnegotiable

⑈0105⑈ 0638⑈ 34 1297 211⑈

4. To Noyes Music Shop for $17.00.

NUMBER 209

01-05 / 638

_____ 19____

PAY TO THE ORDER OF _____ $ _____

_____ DOLLARS

nonnegotiable

CITIZEN'S BANK

Barrington, Oklahoma

⑈0638⑈ 0105⑈ 34 1297 211⑈

Writing Checks

Make out the checks here for the amounts shown. Use today's date and sign your own name.

1. To Hammer Hardware for $21.50.

PLAINS BANK
PUTNEY, VERMONT

Number 299
13-02
1074

_____ 19 ____

PAY TO
THE ORDER OF _____ $ _____

_____ DOLLARS
nonnegotiable

⑆1074⑈ 1302⑆ 85⑉9361⑈0

2. To Cyclops TV Repair for $43.07.

NATIONAL BANK *MERLIN, OREGON*
& TRUST COMPANY

NUMBER
10-74   257
1302

_____ 19 ____

PAY TO
THE ORDER OF _____ $ _____

_____ DOLLARS
nonnegotiable

⑆1302⑈1074⑆ 85⑉9361⑈0

3. To Grand Garage for $98.13.

HOPEWELL BANK
HOPEWELL, IOWA

NUMBER 112
01-16
815

19 _____

PAY TO THE ORDER OF _____ $ _____

_____ DOLLARS
nonnegotiable

_____

⑆0815 ⑈ 0116⑉ 49 378 ⑈ 0⑈

4. To Baron Flying School for $156.50.

ERNEY BANK & TRUST COMPANY
ERNEY, NORTH DAKOTA

Number 300
05-01
638

19 _____

PAY TO
THE ORDER OF _____ $ _____

_____ DOLLARS
nonnegotiable

_____

⑆0638 ⑈ 0501⑉ 34 1297 2⑈

5. To Penelope Weaving, Inc. for $14.40.

BOWDIE NATIONAL BANK
BOWDIE, CALIFORNIA

Number
283
02-41
938

19 _____

Pay to
the Order of _____ $ _____

_____ DOLLARS
nonnegotiable

_____

⑆0938 ⑈ 0241⑉ 618 ⑈51⑈06

The Checkbook Register

Carla started her checking account by putting $200.00 in the bank. Any money that is put into a bank account is called a *deposit*. This deposit was also her *balance*—the total amount of money she had in her account. Carla made a record of the deposit in her checkbook register. She used the checkbook register to keep a record of her deposits and a record of the checks she wrote.

Here is how Carla recorded her first deposit in her checkbook register.

CHECK NO.	DATE	CHECKS ISSUED TO OR DESCRIPTION OF DEPOSIT	AMOUNT OF CHECK	√	AMOUNT OF DEPOSIT	BALANCE	
—	10/7	deposit			200 00	200	00

The first check Carla wrote was for $14.45. She wrote this check to Runaway Sports Shop to pay for two Frisbees. Carla made a record of the check in her checkbook register. Here are the things she recorded:

- the number of the check
- the date she wrote the check
- to whom she wrote the check
- the reason for the check
- the amount of the check
- the new balance in the account

Here is how Carla recorded her check to Runaway Sports Shop. To find the new balance, Carla subtracted the amount of the check from the old balance.

CHECK NO.	DATE	CHECKS ISSUED TO OR DESCRIPTION OF DEPOSIT	AMOUNT OF CHECK	√	AMOUNT OF DEPOSIT	BALANCE	
—	10/7	deposit			200 00	200	00
101	10/15	Runaway Sports Shop (Frisbees)	14 45			185	55

The second check Carla wrote was for $9.67. Here is how she recorded the check in her checkbook register.

CHECK NO.	DATE	CHECKS ISSUED TO OR DESCRIPTION OF DEPOSIT	AMOUNT OF CHECK		√	AMOUNT OF DEPOSIT		BALANCE	
–	10/7	deposit				200	00	200	00
101	10/15	Runaway Sports Shop (Frisbees)	14	45				185	55
102	10/18	Noyes Music Shop (record)	9	67				175	88

Carla wrote check Number 103, shown below, to pay for new clothes. Make a record of this check in Carla's checkbook register above. To find the new balance, subtract the amount of the check from the old balance.

UNITED NATIONAL BANK
Hometown, Illinois

NUMBER 103
09-38
241

October 20 19 89

PAY TO THE ORDER OF _Gant's Department Store_ $ 74 30

———————— Seventy-four and 30/100 ———————— DOLLARS
nonnegotiable

Carla Williams
7294 West Street
Hometown, Illinois

Carla Williams

⑊0241⑊ 0938⑊ 618⑊51⑊06

Carla wrote check Number 104 on October 28. She wrote the check to the *Good News Paper* to pay for her newspaper. The amount of the check was $6.25. Make a record of this check in her checkbook register above. Then figure the new balance.

NAME _____

Balancing the Register

Complete this checkbook register. Figure the new
balance for each check written and each deposit made.
Remember to *subtract* checks and *add* deposits.

CHECK NO.	DATE	CHECKS ISSUED TO OR DESCRIPTION OF DEPOSIT	AMOUNT OF CHECK	√	AMOUNT OF DEPOSIT	BALANCE 427 35
201	4/12	Ma Bell (telephone)	16 24			
202	4/14	IRS (taxes)	58 17			
203	4/14	Maze Dept. Store (clothes)	55 98			
204	4/15	Entropy Power Co. (electric bill)	30 25			
—	4/15	deposit			400 00	
205	4/22	Sundial Jewelers (watch repair)	25 00			
206	4/25	Best Supermarket (groceries)	51 08			
207	4/27	Poore Circulation Co. (magazine)	14 50			
—	4/29	deposit			300 00	
208	5/1	Hillcrest Apartments (rent)	575 00			
209	5/1	Eastside Savings & Loan (savings)	150 00			
210	5/5	OPEC Oil (gasoline)	47 25			
211	5/8	Sparkle Cleaners (dry cleaning)	8 53			
212	5/10	Franklin Drugstore (vitamins)	6 76			

Complete this checkbook register. Figure the new balance for each check written and each deposit made. Remember to *subtract* checks and *add* deposits.

CHECK NO.	DATE	CHECKS ISSUED TO OR DESCRIPTION OF DEPOSIT	AMOUNT OF CHECK		√	AMOUNT OF DEPOSIT		BALANCE 606	71
578	1/26	Fiesta Foods (groceries)	69	89					
579	1/27	Ernie's Garage (oil change)	15	95					
580	1/27	Dr. Yankard (dentist bill)	42	50					
—	1/31	deposit				289	75		
581	2/1	Crown Colony (rent)	545	00					
582	2/5	Longlife Insurance (insurance)	37	85					
583	2/10	Hoopla Theater (tickets)	26	50					
584	2/14	Franklin Florist (flowers)	21	95					
585	2/14	The White Horse (dinner)	32	75					
—	2/15	deposit				425	00		
586	2/18	Great Bank (credit card payment)	78	18					
587	2/18	Ford Motors (car payment)	196	95					
588	2/20	Fiesta Foods (groceries)	25	98					
—	2/28	deposit				375	00		

Checks and Balances

Make a record of these checks and add the deposit in the checkbook register below. Write in all the facts shown. Figure the new balance after each check or deposit. The first one has been done for you.

1. Check Number 105. December 1. To Snow Plow, Inc. For snow blowing. $20.00.

2. Check Number 106. December 3. To Tip Toe Shoes. For boots. $64.79.

3. Check Number 107. December 8. To Franklin Electric. For electricity bill. $37.23.

4. Check Number 108. December 12. To Super Oil Company. For gas. $59.08.

5. Deposit. December 15. $315.00.

6. Check Number 109. December 16. To Noel Evergreens. For Christmas tree. $34.50.

7. Check Number 110. December 20. To Fiesta Foods. For groceries. $73.86.

8. Check Number 111. December 23. To United Charge-It-All. For credit card. $229.99.

CHECK NO.	DATE	CHECKS ISSUED TO OR DESCRIPTION OF DEPOSIT	AMOUNT OF CHECK		√	AMOUNT OF DEPOSIT		BALANCE 231	07
105	12/1	Snow Plow, Inc. (snow blowing)	20	00				211	07

Make a record of these checks and deposits in the
checkbook register below. Write in all the facts shown.
Figure the new balance after each check or deposit.

1. Check Number 238. July 9. To Dr. Murphy.
   For checkup. $50.00.

2. Deposit. July 11. $465.71.

3. Check Number 239. July 12. To North Savings & Loan.
   For savings. $100.00.

4. Check Number 240. July 13. To Trust Insurance.
   For car insurance. $162.60.

5. Check Number 241. July 15. To Handy Hardware.
   For paint. $47.85.

6. Check Number 242. July 20. To *Good News Paper*.
   For newspapers. $6.25.

7. Deposit. July 25. $45.00.

8. Check Number 243. July 30. To Lender's Bank.
   For car payment. $210.15.

CHECK NO.	DATE	CHECKS ISSUED TO OR DESCRIPTION OF DEPOSIT	AMOUNT OF CHECK		√	AMOUNT OF DEPOSIT		BALANCE 84	30

Deposit Slips

Carla filled out a deposit slip each time she put money in her checking account. Each deposit slip had her name, address, and account number printed on it.

On October 31, Carla deposited the cash and checks shown below. Here is how she filled out her deposit slip.

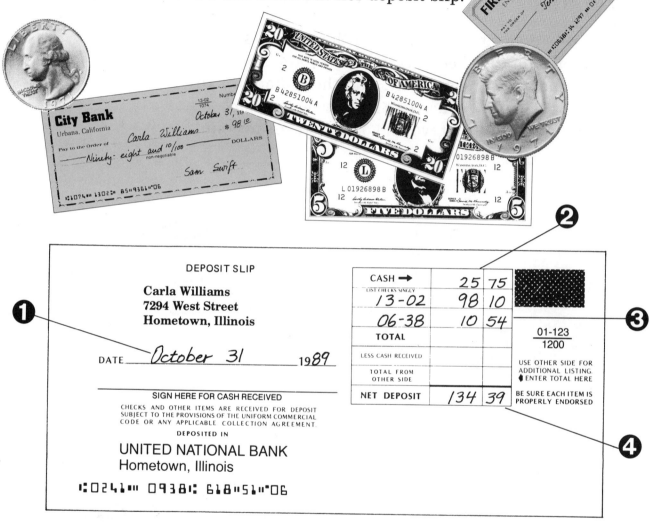

❶ She wrote the date she made the deposit.

❷ She wrote the total amount of cash here.

❸ She listed her checks separately here. On the left, she wrote the bank number (only the number above the line) of each check she deposited. Then she wrote the amount of each check.

❹ She added the amounts of the cash and the checks to get her total deposit.

Steve Pollard deposited the checks shown below. But he didn't put the whole amount into his account. Instead he wanted part of the amount back in cash.

To get the cash back, he had to sign his name on the deposit slip. He also had to write the amount he wanted, $10.00, in the space marked *less cash received*. He then subtracted $10.00 from the total of the checks.

Fill in Steve's deposit slip below with the amounts of money shown on the checks. Use today's date. Remember to list each check separately. Add these amounts to get the *total*. Then subtract the $10.00 to get the *net deposit*—the amount that would go into Steve's account.

DEPOSIT SLIP	CASH ➡		
Steve Pollard **20 Parkway East** **Oaklawn, Illinois**	LIST CHECKS SINGLY		
	TOTAL		
DATE_____ 19___	LESS CASH RECEIVED		
Steve Pollard	TOTAL FROM OTHER SIDE		
SIGN HERE FOR CASH RECEIVED	**NET DEPOSIT**		

01-123 / 1200

USE OTHER SIDE FOR ADDITIONAL LISTING.
◆ ENTER TOTAL HERE

BE SURE EACH ITEM IS PROPERLY ENDORSED

CHECKS AND OTHER ITEMS ARE RECEIVED FOR DEPOSIT SUBJECT TO THE PROVISIONS OF THE UNIFORM COMMERCIAL CODE OR ANY APPLICABLE COLLECTION AGREEMENT.

DEPOSITED IN

UNITED NATIONAL BANK
Hometown, Illinois

⑆0241⑆ 0938⑆ 518⑈61⑈

Deposit Slips

Fill out the deposit slip below for the amounts of money shown here. Use today's date. Remember to list each check separately. Then add to find the total amount of the deposit.

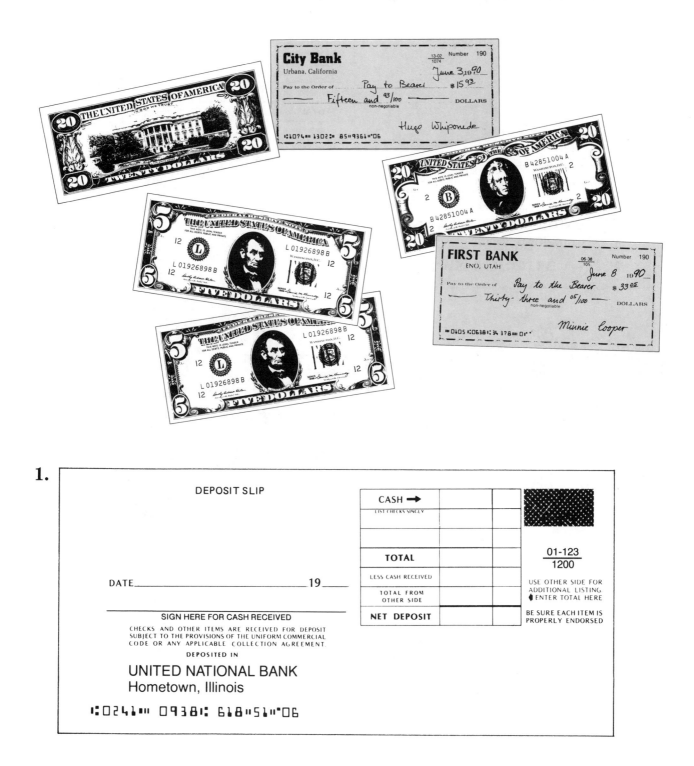

City Bank
Urbana, California

13-02 / 1074   Number 190

SEPT 10 19 89

Pay to the Order of PAY TO BEARER $16.52

SIXTEEN AND 52/100 ____ DOLLARS
non-negotiable

⑈1074⑈⑈0815 ⑈ 85⑈93⑈

FIRST BANK
ENO, UTAH

06-38 / 105   NUMBER

Sept. 15 19

PAY TO THE ORDER OF Pay to bearer $105.80

One hundred and five and 80/100 ____ Dollars
non-negotiable

Paul Quinn

⑈0105 ⑈⑈0638⑈ 34 4297 ⑈ 0⑈

2.

DEPOSIT SLIP

CASH ➡		
LIST CHECKS SINGLY		
TOTAL		
LESS CASH RECEIVED		
TOTAL FROM OTHER SIDE		
NET DEPOSIT		

01-123
1200

DATE_____ 19____

SIGN HERE FOR CASH RECEIVED

CHECKS AND OTHER ITEMS ARE RECEIVED FOR DEPOSIT
SUBJECT TO THE PROVISIONS OF THE UNIFORM COMMERCIAL
CODE OR ANY APPLICABLE COLLECTION AGREEMENT.

DEPOSITED IN

UNITED NATIONAL BANK
Hometown, Illinois

⑈0241⑈ 0938⑈ 618⑈51⑈06

USE OTHER SIDE FOR
ADDITIONAL LISTING.
◆ENTER TOTAL HERE

BE SURE EACH ITEM IS
PROPERLY ENDORSED

FIRST BANK
ENO, UTAH

06-38 / 105   NUMBER 225

PAY TO THE ORDER OF Pay to the Bearer   Nov. 6 19 89

Three hundred thirteen and 05/100 ____ $313.05
non-negotiable

William Nagel

⑈0105 ⑈⑈0638⑈ 34 4297 ⑈ 0⑈

City Bank
Urbana, California

13-02 / 1074

Nov 2 19

Pay to the Order of Pay to bearer $15.98

Fifteen and 98/100 ____ DOLLARS
non-negotiable

⑈1074⑈13 49 85⑈9361⑈06

3.

DEPOSIT SLIP

CASH ➡		
LIST CHECKS SINGLY		
TOTAL		
LESS CASH RECEIVED		
TOTAL FROM OTHER SIDE		
NET DEPOSIT		

01-123
1200

DATE_____ 19____

SIGN HERE FOR CASH RECEIVED

CHECKS AND OTHER ITEMS ARE RECEIVED FOR DEPOSIT
SUBJECT TO THE PROVISIONS OF THE UNIFORM COMMERCIAL
CODE OR ANY APPLICABLE COLLECTION AGREEMENT.

DEPOSITED IN

UNITED NATIONAL BANK
Hometown, Illinois

⑈0241⑈ 0938⑈ 618⑈51⑈06

USE OTHER SIDE FOR
ADDITIONAL LISTING.
◆ENTER TOTAL HERE

BE SURE EACH ITEM IS
PROPERLY ENDORSED

A Savings Account

Soon after Steve Pollard started working at The Sandwich House, he opened a savings account at Farmers First Bank.

Each time Steve wanted to put money into his account, he had to fill out a deposit slip. On November 21, Steve deposited the cash and checks shown below. Here is how he filled out his deposit slip.

SAVINGS DEPOSIT

❶ Account Number _04-008074-3_   DATE _11_ / _21_ / _89_ **❸**
MO   DAY   YR

❷ Name _Steve Pollard_

Deposited with
Farmers First Bank

LIST CHECKS BY BANK NUMBER	DOLLARS	CENTS	
CURRENCY	30	00	**❹**
COIN	1	00	
CHECKS *13-02*	16	52	**❺**
09-38	25	00	
CHECKS FROM OTHER SIDE			◄ IF MORE THAN 2 CHECKS LIST ON REVERSE SIDE AND ENTER TOTAL HERE
SUBTOTAL			
LESS CASH RECEIVED			
DEPOSIT TOTAL	72	52	**❻**

❶ He wrote his savings account number here.

❷ He printed his name here.

❸ He wrote the date he made the deposit.

❹ He listed the amount of bills and coins separately here.

❺ He listed the bank number and the amount of each check here.

❻ He added the amounts of the cash and the checks to get his total deposit.

Fill out the deposit slips below with the amounts of
money shown. Use today's date and print your own
name.

1.

SAVINGS DEPOSIT

Account Number ___07- 100099-9___   DATE ___/___/___
　　　　　　　　　　　　　　　　　　　　　　MO / DAY / YR

Name _____

LIST CHECKS BY BANK NUMBER	DOLLARS	CENTS	
CURRENCY			
COIN			
CHECKS			
CHECKS FROM OTHER SIDE			◄ IF MORE THAN 2 CHECKS LIST ON REVERSE SIDE AND ENTER TOTAL HERE
SUBTOTAL			
LESS CASH RECEIVED			
DEPOSIT TOTAL			

Deposited with
Farmers First Bank

2.

SAVINGS DEPOSIT

Account Number ___01 - 50666 - 8___   DATE ___/___/___
　　　　　　　　　　　　　　　　　　　　　　MO / DAY / YR

Name _____

LIST CHECKS BY BANK NUMBER	DOLLARS	CENTS	
CURRENCY			
COIN			
CHECKS			
CHECKS FROM OTHER SIDE			◄ IF MORE THAN 2 CHECKS LIST ON REVERSE SIDE AND ENTER TOTAL HERE
SUBTOTAL			
LESS CASH RECEIVED			
DEPOSIT TOTAL			

Deposited with
Farmers First Bank

Steve's Passbook

When Steve opened his savings account, the bank gave him a passbook. The passbook showed how much money Steve had in his account. Each time Steve put money into his account, the bank teller recorded the deposit in Steve's passbook. Sometimes Steve took some money out of his account. This is called making a *withdrawal*. Each time Steve made a withdrawal, the teller recorded it in Steve's passbook.

The bank paid Steve money for keeping his savings there. This money is called *interest*. The amount of interest Steve earned on his savings was also shown in his passbook. The last column in the passbook showed the balance in the account after each deposit or withdrawal.

DATE	WITHDRAWAL	DEPOSIT	INTEREST	BALANCE
11/21		72.52		72.52
11/30		25.00	.12	97.64
12/15	35.00			62.64
12/21	40.00		.31	22.95
1/13		30.00		52.95
1/31		25.00	.17	78.12

Here is how Steve's passbook looked after three months. Study the passbook and answer the questions below.

1. How much money did Steve have in his account after making his second deposit? _____

2. On December 21, what was Steve's balance? _____

3. How much did Steve withdraw *altogether* from his account? _____

4. How much did Steve deposit *altogether* in his account? _____

5. How much interest did Steve earn on his savings? _____

Bob's Balance

Bob Sommers also had a savings account at Farmers First Bank. Here are the entries for his account for four months.

DATE	ENTRY	AMOUNT
3/2	Deposit	38.50
3/20	Deposit	25.00
3/30	Deposit	45.00
3/30	Interest	.12
4/6	Withdrawal	35.50
4/21	Deposit	23.75

DATE	ENTRY	AMOUNT
5/4	Deposit	30.00
5/25	Withdrawal	25.00
5/31	Interest	1.17
6/1	Deposit	36.50
6/15	Deposit	23.50
6/29	Withdrawal	50.00

Write each entry in the correct column of Bob's passbook. Remember to figure the new balance after each entry.

DATE	WITHDRAWAL	DEPOSIT	INTEREST	BALANCE
3/2		38.50		38.50

On June 29, what was Bob's balance? _____

What's the Balance?

Write each of these ent ·es in the correct column of the
passbook below. Figure he new balance after each entry.

DATE	ENTRY	AMOUNT
6/1	Deposit	50.00
6/15	Deposit	38.50
7/6	Withdrawal	20.00
7/18	Deposit	52.25
7/31	Interest	.85
8/10	Deposit	46.75
8/18	Deposit	15.00
8/27	Withdrawal	35.00

DATE	ENTRY	AMOUNT
9/7	Deposit	55.00
9/21	Deposit	32.75
9/28	Interest	1.60
10/1	Deposit	48.50
10/12	Withdrawal	50.00
10/16	Deposit	25.00
11/9	Withdrawal	37.00
11/23	Deposit	20.00

DATE	WITHDRAWAL	DEPOSIT	INTEREST	BALANCE
6/1		50.00		50.00

On November 23, what was the balance in this account? _____

Saving Wheel

Figure each person's balance by adding deposits and interest and by subtracting withdrawals. Start with the deposit in the center of the wheel. Then work outward toward the rim, filling in the blanks as you go. (As you work toward the rim, each answer is part of the next problem.)

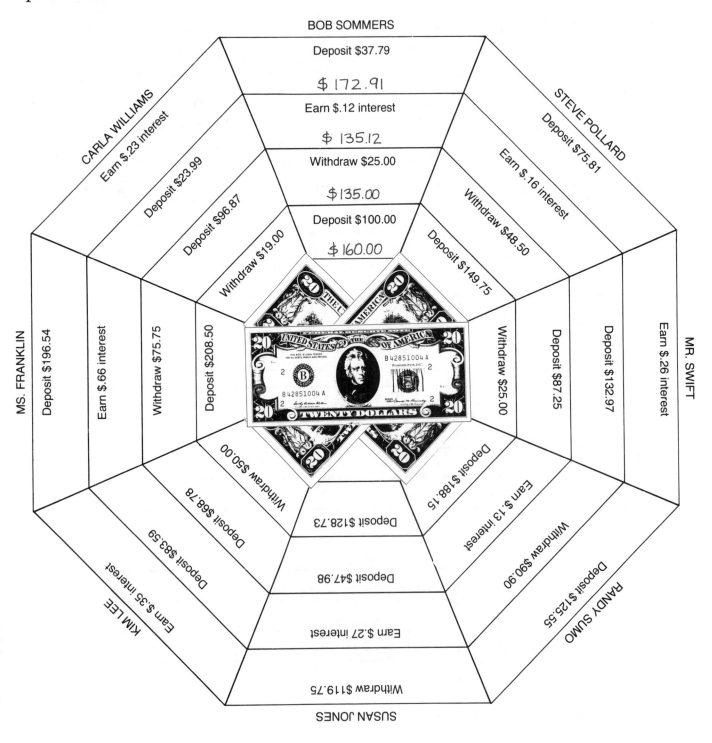

BOB SOMMERS
Deposit $37.79
$ 172.91
Earn $.12 interest
$ 135.12
Withdraw $25.00
$ 135.00
Deposit $100.00
$ 160.00

CARLA WILLIAMS
Earn $.23 interest
Deposit $23.99
Deposit $96.87
Withdraw $19.00

STEVE POLLARD
Deposit $75.81
Earn $.16 interest
Withdraw $48.50
Deposit $149.75

MS. FRANKLIN
Deposit $196.54
Earn $.66 interest
Withdraw $75.75
Deposit $208.50

MR. SWIFT
Earn $.26 interest
Deposit $132.97
Deposit $87.25
Withdraw $25.00

KIM LEE
Earn $.35 interest
Deposit $83.59
Deposit $68.78
Withdraw $50.00

SUSAN JONES
Withdraw $119.75
Earn $.27 interest
Deposit $47.98
Deposit $128.73

RANDY SUMO
Deposit $125.55
Withdraw $90.90
Earn $.13 interest
Deposit $188.15

Automated Teller Machines

Sometimes Carla wanted to go to the bank in the evenings or on the weekend. But her bank wasn't open during these times. Carla discovered that her bank had an *automated teller machine (ATM)*. An ATM is a computer-operated machine that does bank *transactions*—usually deposits and withdrawals. The ATM at Carla's bank operated every day from six in the morning until midnight. Some ATMs are open 24 hours a day.

Here are some reasons for using an ATM.

- You can get money from your account, even when the bank is closed.
- The lines at the teller machines are often shorter than inside the bank.
- At some stores and gas stations, you can use your ATM card instead of writing a check. The amount is subtracted from your account.

Carla filled out an application for an ATM card. She soon got the card in the mail. The card had a number printed on the front. It also came with a four-number code that was not on the card. This code was for Carla to use when starting a transaction. No one but Carla would know the code. This way, only Carla could use her card to take money from her account.

UNB United National Bank **E-Z Teller**

9 4 4 6 7 2 0

Dates 12/89 to 12/90

CARLA L. WILLIAMS

Here are some things to remember about using an ATM card.

- You can only withdraw money that you have in your account.
- You should never deposit cash.
- Most banks have a limit on the amount you can withdraw each day.

One evening Carla went to the bank. She wanted to take some money from her checking account. She started the transaction by putting her ATM card into the slot on the teller machine. Then she pressed the number buttons to enter her code.

Instructions appeared on the screen, telling Carla which buttons to push for each step. The buttons she pushed told the machine how much money she wanted, and from which account she wanted to take it. The money came out through another slot on the machine. Carla took the money, then pushed a button to show that she was finished. Her card was returned, and she received a record of her transaction.

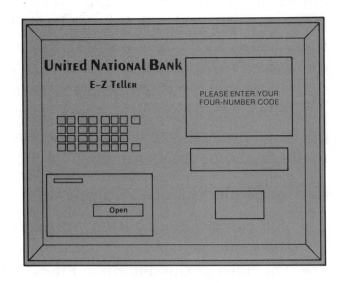

❶E-Z Teller
United National Bank
TRANSACTION RECORD—PLEASE SAVE

ATM Number	Card Number	❹ Transaction
❷ 0838	❸ 9446720	Withdrawal
❺ **Amount**	**Date**	❻ **Time**
$40.00	12/22/89	7:20 p.m.
❼ **Account**	**Location**	**New Balance**
Checking #618-51-06	Hometown Center ❽	$308.56 ❾

THANK YOU

Transaction Records

These are the items that appeared on Carla's transaction record.

1 *E-Z Teller.* This is the name United National Bank uses for its automated teller machines.

2 *The ATM number.* Each United National Bank has its own number.

3 *The number on Carla's ATM card.* This number is different from Carla's account number, and it's different from her four-number code.

4 *The type of transaction.* Carla made a withdrawal from her account.

5 *The amount of money Carla took out.* Each bank has a limit on the amount a person can take out each day. The limit at Carla's bank is $200.00.

6 *The date and time* that the transaction took place.

7 *Carla's checking account number.*

8 *The place the transaction was made.* This United National Bank is in the Hometown Shopping Center.

9 *Carla's balance.* This is the amount in her account *after* she took out the $40.00.

Using the transaction record, Carla made a record of her withdrawal in her checkbook register.

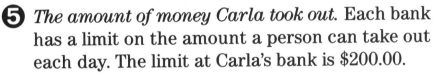

CHECK NO.	DATE	DESCRIPTION OF CHECK, DEPOSIT, OR WITHDRAWAL	AMOUNT—CHECK OR WITHDRAWAL	√	AMOUNT OF DEPOSIT	BALANCE	
						348	56
—	12/22	ATM Withdrawal	40 00			308	56

Steve Pollard made the following transaction at his bank's automated teller machine. Study the transaction record and answer the questions below.

SAVE TIME TELLER
Farmers First Bank

TRANSACTION: Deposit   LOCATION: Gallee

CARD NUMBER: 0276494   ACCOUNT: Savings #002-9189

ATM NUMBER: 0589   AMOUNT: $105.00

DATE: 1/16/90   NEW BALANCE: $465.23

TIME: 5:15 p.m.

1. What type of transaction did Steve make, a deposit or a withdrawal? _____

2. At what bank does Steve have this account?

   _____

3. What is Steve's account number? _____

4. How much money did Steve deposit or withdraw?

   _____

5. What was Steve's balance after this transaction?

   _____

Record this transaction in Steve's passbook.

DATE	WITHDRAWAL	DEPOSIT	INTEREST	BALANCE
1/2	25.00			360.23

Cashing and Depositing Checks

Before Carla cashed a check at the bank, she had to *endorse* it. This means she signed her name on the back of the check. She usually waited until she got to the bank to endorse a check. She knew that if someone found her endorsed check, that person could try to cash it.

When Carla endorsed a check, she signed her name just the way it was written on the front. By endorsing it this way, she could *cash* the check— take the money for it. Or she could deposit part or all of it into her account.

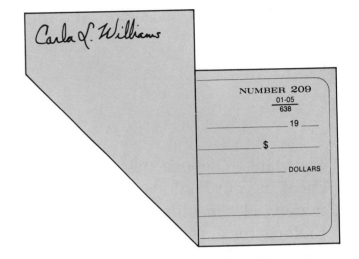

Sometimes Carla needed to endorse a check when she wasn't going to the bank. Then she wrote the words "For Deposit Only" in addition to her name. This way, no one else could cash her check.

These are some of the times you would endorse a check with "For Deposit Only":

- When you deposit a check without asking for any cash back.
- When you mail a deposit to your bank.
- When you deposit a check at an automated teller machine.
- When you give a check to someone else to deposit for you.

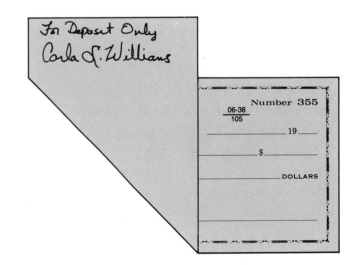

Carla had a check that was made out to her for $10.00. She wanted to give the $10.00 to her friend, Mary Rossi. Since Mary was going to the bank, Carla endorsed the check so Mary could cash it. To do this, she wrote the words "Pay to the Order of" followed by Mary's name. When Mary cashed the check, she endorsed it with her own name.

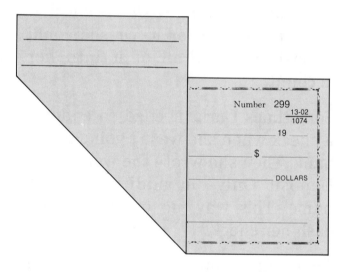

You want to deposit this check at your bank's automated teller machine. Endorse the check with your name and the words "For Deposit Only."

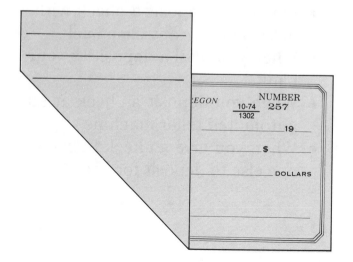

You are giving this check to Peter Rock. Endorse it so he can cash it. Use the words "Pay to the Order of."

The Bank Statement

Once a month Carla's bank sent her a *bank statement*. The statement listed all the deposits and withdrawals Carla made during the past month. It also listed the check numbers and amounts of all the checks she wrote. The statement showed how much money she had at the beginning and at the end of the month.

Here is the bank statement Carla got for the month of September.

❶ Closing date. All the transactions listed on this statement were made before this date.

❷ Beginning balance. Carla had $302.75 at the beginning of the month covered by this statement.

❸ Checks. This part lists all the checks Carla wrote that have been canceled. The checks are listed in order by check number.

❹ Other charges. This part lists all the withdrawals Carla made at the bank's ATM. It also lists the service charge for Carla's checking account. The bank subtracts this amount from Carla's account once a month.

❺ Deposits. The statement lists the deposits Carla made, both at the ATM and at the bank.

UNITED NATIONAL BANK
Hometown, Illinois 62000
STATEMENT

Carla Williams
7294 West Street
Hometown, Illinois 62000

Closing Date: 4/30/90
Beginning Balance: 302.75

CHECKING ACCOUNT NUMBER: 618-51-06

CHECKS

Check Number	Date Paid	Amount	Check Number	Date Paid	Amount
298	4/3/90	17.84			
299	4/8/90	25.19			
300	4/9/90	10.50			
301	4/14/90	55.58			
*					
303	4/26/90	115.00			

OTHER CHARGES

	Date	Amount
WITHDRAWAL ATM 0838	4/12/90	40.00
SERVICE CHARGE	4/30/90	5.00

DEPOSITS

Date	Amount	Date	Amount
4/7/90	104.93		
4/21/90	209.86		

Ending Balance: 357.11

❻ Ending balance. After all her deposits, withdrawals, and service charges were counted, Carla had $357.11 in her account.

Here is another bank statement. Study it and answer the question below.

1. How many canceled checks are listed?

_____

2. What was the amount of the deposit made on 11/10/89?

_____

3. What is the check number of the check that has not yet been canceled? (Space is marked with a star.)

_____

4. On what date did Sandy Jackson withdraw $40.00 from the ATM?

_____

5. How much was Sandy's ending balance on this statement?

_____

6. How much is the service charge for Sandy's checking account?

_____

7. What is the total amount of all the canceled checks on this statement?

_____

UNITED NATIONAL BANK
Hometown, Illinois 62000

STATEMENT

Sandy Jackson
392 Aspen Street
Antioch, Illinois 62001

Closing Date: 11/30/89
Beginning Balance: $453.88

CHECKING ACCOUNT NUMBER: 009-73852

CHECKS

Check Number	Date Paid	Amount	Check Number	Date Paid	Amount
398	11/2/89	63.85			
399	11/5/89	206.37			
400	11/10/89	10.00			
401	11/16/89	26.54			
*					
403	11/20/89	140.00			
404	11/28/89	73.79			

OTHER CHARGES	Date	Amount
WITHDRAWAL ATM 9004	11/25/89	40.00
WITHDRAWAL ATM 0839	11/29/89	20.00
SERVICE CHARGE	11/30/89	5.00

DEPOSITS				
Date	Amount		Date	Amount
11/10/89	204.06			
11/24/89	265.98			

Ending Balance: $338.37

Balancing the Checking Account

The balance in Carla's checkbook register did not usually match the closing balance on her statement. Carla knew why. She always subtracted the amount of each check from her balance. But not every check she wrote was canceled before the statement's closing date.

If a check wasn't canceled, the bank didn't subtract it from Carla's balance. So the balance on the statement was different than the balance in the register.

Carla still used the information on her statement. She used it to make sure both she and the bank recorded everything correctly. This is called *balancing the account.*

Balancing a checking account is not easy. It takes a lot of practice. These two pages will give you an idea of how it can be done.

When Carla balanced her checking account, she first looked at her checkbook register. She put a mark next to each check that appeared on her statement. She made sure the amount of each check in her checkbook register matched the amount on her statement.

Two checks in Carla's register were not on her statement. Checks not listed on the statement are called *outstanding checks.*

CHECK NO	DATE	CHECKS ISSUED TO OR DESCRIPTION OF DEPOSIT	AMOUNT OF CHECK		√	AMOUNT OF DEPOSIT		BALANCE 302 75	
298	4/3	Graham Telephone	17	84	√			284	91
	4/7	deposit				104	93	389	84
299	4/8	Famished Foods	25	19	√			364	65
300	4/9	King's Way	10	50	√			354	15
	4/12	ATM Withdrawal	40	00	√			314	15
301	4/14	Bliss Jewelers	55	58	√			258	57
302	4/20	General Oil (charge card payment)	32	60				225	97
	4/21	deposit				209	86	435	83
303	4/26	Rachel Isaacson	115	00	√			320	83
304	5/2	Solestruck Shoe Store	36	42				288	19
	5/5	deposit				89	93	378	02
	4/30	service charge	5	00	√			373	02

On the back of Carla's checkbook register, there were directions to help her balance her account. Here is how Carla followed the directions.

❶ Carla wrote down the check number and amount of each outstanding check. Then she added to get the total. (Remember, outstanding checks appeared in her register, but not on the statement.)

❷ Carla wrote down the ending balance, shown on the front of the statement.

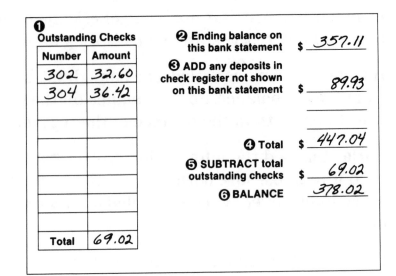

❶ Outstanding Checks	
Number	**Amount**
302	32.60
304	36.42
Total	69.02

❷ Ending balance on this bank statement $ 357.11

❸ ADD any deposits in check register not shown on this bank statement $ 89.93

❹ Total $ 447.04

❺ SUBTRACT total outstanding checks $ 69.02

❻ BALANCE 378.02

❸ In Carla's checkbook register, there was one deposit that didn't appear on the statement. She had made this deposit after the closing date of the statement. Carla wrote down the amount of the deposit in the space.

❹ Carla added together the ending balance and the deposit.

❺ She subtracted the total of the outstanding checks.

❻ The amount on this line matched the balance in Carla's checkbook register. Carla's checking account *balanced.*

You won't always be able to balance your account the first time you try. You may need to check for mistakes in addition or subtraction. Sometimes it still won't balance, even after you try many times. Then you may need to ask the bank to help you.

Percent

The money in your savings account is used by the bank (or savings and loan institution) to make loans to other people. In return for using your money, the bank pays you interest. The interest you get is a *percent* (%) of the amount in your account.

Percent is a way of figuring in *hundredths*.
1% (one percent) is $\frac{1}{100}$ (one-hundredth).
2% (two percent) is $\frac{2}{100}$ (two-hundredths).
3% (three percent) is $\frac{3}{100}$ (three-hundredths).

Percent can also be written as a decimal fraction. For example, to change 3% to a decimal fraction, drop the percent sign (%) and put the decimal point *two places to the left*.

$3\% = .03$ (three-hundredths)

To change 12% to a decimal fraction, follow the same steps. Drop the percent sign and put the decimal point two places to the left.

$12\% = .12$ (twelve-hundredths)

Change these percents to decimal fractions. Drop the percent sign and put the decimal point two places to the left.

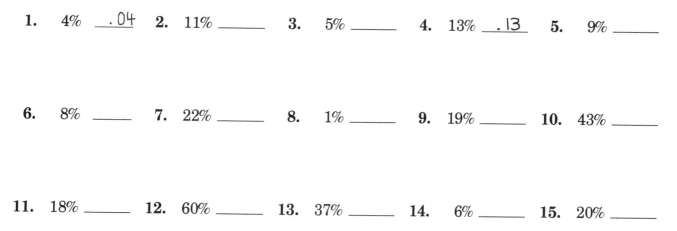

1. 4% $.04$ 2. 11% _____   3. 5% _____   4. 13% $.13$ 5. 9% _____

6. 8% _____   7. 22% _____   8. 1% _____   9. 19% _____   10. 43% _____

11. 18% _____   12. 60% _____   13. 37% _____   14. 6% _____   15. 20% _____

Interest

Let's say you have $100.00 in your savings account. And let's say the bank pays 5% interest a year. How much interest will you get? To find out, first change 5% to a decimal fraction. Then multiply the amount in your account by the decimal fraction.

$$5\% = .05$$

$$
\begin{array}{r}
\$100.00 \\
\times\ .05 \\
\hline
\$5.0000
\end{array}
$$

There are four numbers after the decimal points in the problem, so the decimal point in the answer goes *four places to the left*.

$$
\begin{array}{r}
^{1\ 2} \\
\$100.00 \\
^{3\ 4} \\
\times\ .05 \\
\hline
^{4\ 3\ 2\ 1} \\
\$5.0000
\end{array}
$$

Cross out the last two zeros. The interest on $100.00 at 5% is $5.00.

$$\$5.00\cancel{00}$$

Figure the interest on each amount here. Change the percent to a decimal fraction. Then multiply the amount by the decimal fraction. Be sure to put the decimal point in the right place in your answers.

1. $200.00 at 5% = $10.00

$$
\begin{array}{r}
\$200.00 \\
\times\ \ \ \ .05 \\
\hline
10.00\cancel{00} \\
\$10.00
\end{array}
$$

2. $100.00 at 6% = _____

3. $300.00 at 7% = _____

4. $200.00 at 6% = _____

5. $400.00 at 7% = _____

6. $300.00 at 5% = _____

7. $100.00 at 7% = _____

8. $200.00 at 7% = _____

9. $300.00 at 6% = _____

NAME _____

Interest

Figure the interest on each amount here. Change the
percent to a decimal fraction. Then multiply the amount
by the decimal fraction. Be sure to put the decimal
point in the right place in your answers.

1. $250.00 at 5% = $12.50

    $ 250.00
    x   .05
    12.50ØØ
    $12.50

2. $350.00 at 7% = _____

3. $200.00 at 6% = _____

4. $150.00 at 5% = _____

5. $400.00 at 8% = _____

6. $125.00 at 6% = _____

7. $250.00 at 6% = _____

8. $325.00 at 7% = _____

9. $175.00 at 5% = _____

10. $420.00 at 8% = _____

11. $280.00 at 5% = _____

12. $440.00 at 9% = _____

13. $134.00 at 5% = _____

14. $268.00 at 6% = _____

15. $166.00 at 7% = _____

16. $324.00 at 5% = _____

17. $239.00 at 8% = _____

18. $109.00 at 6% = _____

Interesting Wheels!

Figure the interest on the amount in the center of each wheel at the interest rates between the spokes.

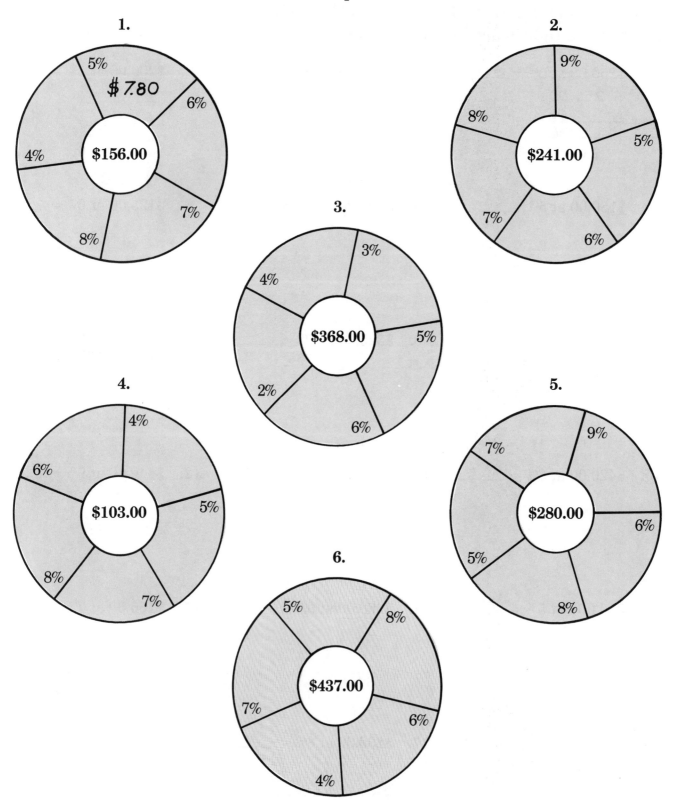

1.

5%
$7.80
6%
4%
$156.00
7%
8%

2.

9%
8%
5%
$241.00
7%
6%

3.

3%
4%
$368.00
5%
2%
6%

4.

4%
6%
5%
$103.00
8%
7%

5.

9%
7%
6%
$280.00
5%
8%

6.

5%
8%
$437.00
7%
6%
4%

A Taxing Job

Steve Pollard liked working at the Sandwich House. He liked meeting people. And, as a waiter, Steve met a lot of people every day.

Part of Steve's job was figuring the total cost of each order. He listed everything that was ordered and added the price of each thing. This amount he wrote in the *subtotal* box of the bill.

Next, Steve figured the sales tax. The sales tax was 6% of the subtotal. He added the sales tax to the subtotal to get the total.

Steve used a table that showed him what 6% of the subtotal was. But he also knew how to figure 6% without using a table. Here is how Steve figured the tax on a subtotal of $3.00.

He changed 6% to a decimal fraction.
   6% = .06

Then he multiplied the subtotal by the fraction. He moved the decimal point *four places to the left* in his answer and dropped the last two zeros.

$3.00 **subtotal** $3.00 **subtotal**
× .06   + .18 **sales tax**
$.18ØØ **sales tax** $3.18 **total**

The sales tax on $3.00 is $.18. Steve added $.18 to the subtotal to get the total—$3.18.

The Sandwich House	
ORDER	COST
1 sandwich Special	$2.15
1 milk	.85
SUBTOTAL	$3.00
SALES TAX	.18
TOTAL	$3.18

What's the Total?

Add the orders on each bill to get the subtotal. Figure 6% sales tax for the subtotal. Add the sales tax to the subtotal to get the total.

1.

The Sandwich House

ORDER	COST
1 cheese sandwich	$ 2.75
1 orange drink	.85
1 pie	1.90
SUBTOTAL	
SALES TAX	
TOTAL	

2.

The Sandwich House

ORDER	COST
1 turkey sandwich	$ 3.20
1 soup	1.75
1 salad	2.20
1 coffee	.75
SUBTOTAL	
SALES TAX	
TOTAL	

3.

The Sandwich House

ORDER	COST
1 chicken sandwich	$ 3.15
1 cheese sandwich	2.75
1 fruit plate	3.50
1 milk	.85
1 orange drink	.85
SUBTOTAL	
SALES TAX	
TOTAL	

4.

The Sandwich House

ORDER	COST
2 ham sandwiches	$ 6.60
1 fruit plate	3.50
1 soup	1.75
3 milks	2.55
1 salad	2.20
2 pie	3.80
SUBTOTAL	
SALES TAX	
TOTAL	

Shopping Quiz

Figure 5% sales tax for each of these things. Add the sales tax to the price to get the total cost.

1.

Price $6.60

Sales tax _____ *.33*

Total _____ *$6.93*

2.

Price $5.00

Sales tax _____

Total _____

3.

Price $3.20

Sales tax _____

Total _____

4.

Price $24.40

Sales tax _____

Total _____

5.

Price $10.00

Sales tax _____

Total _____

6.

Price $18.00

Sales tax _____

Total _____

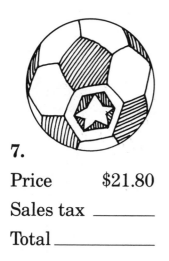

7.

Price $21.80

Sales tax _____

Total _____

8.

Price $31.20

Sales tax _____

Total _____

9.

Price $19.40

Sales tax _____

Total _____

10.

Price $20.80

Sales tax _____

Total _____

11.

Price $21.80

Sales tax _____

Total _____

12.

Price $48.00

Sales tax _____

Total _____

13.

Price $22.20

Sales tax _____

Total _____

14.

Price $26.80

Sales tax _____

Total _____

15.

Price $52.60

Sales tax _____

Total _____

16.

Price $76.80

Sales tax _____

Total _____

17.

Price $54.40

Sales tax _____

Total _____

18.

Price $57.80

Sales tax _____

Total _____

19.

Price $61.00

Sales tax _____

Total _____

20.

Price $98.50

Sales tax _____

Total _____

21.

Price $190.80

Sales tax _____

Total _____

25% Off!

Gant's Department Store was going to have a sale. Everything on sale would be 25% off the regular price.

If the regular price of a hair dryer was $15.95, what would the sale price be?

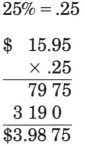

First change 25% to a decimal fraction.

$25\% = .25$

Then multiply the regular price by the fraction. Move the decimal point four places to the left.

```
$  15.95
×  .25
────────
   79 75
 3 19 0
────────
$3.98 75
```

Now round off your answer. Since the third number to the right of the decimal point (7) is greater than 5, round off to the next highest cent. Drop the last two numbers, and your answer is $3.99.

$3.98\cancel{75} = \$3.99$

Subtract $3.99 from the regular price to get the sale price. The sale price of the hair dryer is $11.96.

```
$15.95
− 3.99
──────
$11.96
```

To round off numbers to the nearest cent, remember: If the third number to the right of the decimal point is 5 or more, round off your answer to the next highest cent. If the third number to the right of the decimal point is less than 5, just drop the last two numbers.

Figure the sale price of each thing below. First find 25% of the regular price. Round off your answer to the nearest cent. Then subtract that amount from the regular price to get the sale price.

1.

Regular $21.25

25% off _____ 5.31

Sale price _____ $15.94

2.

Regular $22.40

25% off _____

Sale price _____

3.

Regular $9.80

25% off _____

Sale price _____

4.

Regular $19.75

25% off _____

Sale price _____

5.

Regular $12.89

25% off _____

Sale price _____

6.

Regular $37.50

25% off _____

Sale price _____

7.

Regular $14.95

25% off _____

Sale price _____

8.

Regular $16.37

25% off _____

Sale price _____

9.

Regular $12.76

25% off _____

Sale price _____

10.

Regular $24.95

25% off _____

Sale price _____

11.

Regular $13.99

25% off _____

Sale price _____

12.

Regular $45.00

25% off _____

Sale price _____

Profit

Here are some of the things sold in Gant's Department Store. The store's cost for each thing is shown. The owner of the store wants to make 28% profit on these things. Figure the selling price of each thing by adding the cost and the profit. Write your answers in the blanks.

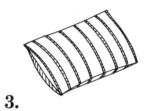

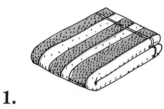

1.
Cost $10.80
28% Profit _3.02_
Price _$13.82_

2.
Cost $6.99
28% Profit _____
Price _____

3.
Cost $12.11
28% Profit _____
Price _____

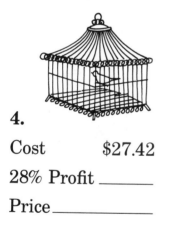
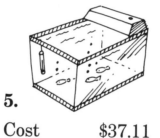
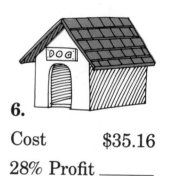

4.
Cost $27.42
28% Profit _____
Price _____

5.
Cost $37.11
28% Profit _____
Price _____

6.
Cost $35.16
28% Profit _____
Price _____

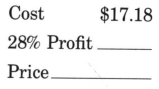

7.
Cost $13.25
28% Profit _____
Price _____

8.
Cost $12.89
28% Profit _____
Price _____

9.
Cost $17.18
28% Profit _____
Price _____

Here are some other things the store sells for different
profit percents. Figure the price of each one.

10.

Cost $36.52

28% Profit _____

Price _____

11.

Cost $23.06

30% Profit _____

Price _____

12.

Cost $25.74

35% Profit _____

Price _____

13.

Cost $61.38

23% Profit _____

Price _____

14.

Cost $21.29

36% Profit _____

Price _____

15.

Cost $20.83

20% Profit _____

Price _____

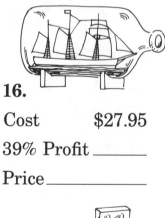

16.

Cost $27.95

39% Profit _____

Price _____

17.

Cost $34.42

20% Profit _____

Price _____

18.

Cost $20.96

24% Profit _____

Price _____

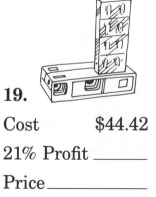

19.

Cost $44.42

21% Profit _____

Price _____

20.

Cost $48.47

34% Profit _____

Price _____

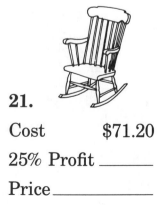

21.

Cost $71.20

25% Profit _____

Price _____

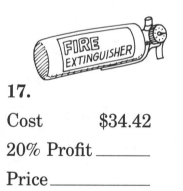

Percent in Action

Find the answers to these percent problems. Show your work below each problem and write the answer in the blank. Be sure to put your decimal point in the right place. Round off your answers to the nearest cent.

1.   5% of $7.52 = _____

2.   9% of $9.87 = _____

3.   6% of $5.14 = _____

4.   5% of $12.70 = _____

5.   7% of $15.32 = _____

6.   8% of $14.99 = _____

7.   12% of $23.33 = _____

8.   11% of $27.09 = _____

9.   18% of $30.76 = _____

10.   8% of $258.00 = _____

11.   9% of $169.00 = _____

12.   7% of $612.00 = _____

13.   25% of $46.98 = _____

14.   20% of $75.55 = _____

15.   45% of $84.32 = _____

The Buying Game

Start with the store's cost of each thing below. Add the store's profit. Subtract the sale amount. Then add the sales tax to get the total cost.

	STORE'S COST	SELLING PRICE	SALE PRICE	TOTAL COST
1.	$20.00	Add 35% profit $ 27.00	Subtract 10% $ 24.30	Add 6% sales tax $ 25.76
2.	$34.00	Add 15% profit	Subtract 12%	Add 5% sales tax
3.	$68.00	Add 18% profit	Subtract 9%	Add 7% sales tax
4.	$119.00	Add 22% profit	Subtract 15%	Add 4% sales tax
5.	$247.00	Add 15% profit	Subtract 5%	Add 6% sales tax

Buying Groceries

Steve Pollard spent part of his earnings on groceries. Sometimes he needed to divide to find out what something cost.

Steve bought ½ dozen eggs. Eggs cost $.98 per dozen. To find the price of ½ dozen, Steve divided by 2—the bottom number of the fraction ½.

```
       $.49  price per ½ dozen
  2 | $.98  price per dozen
       8
       18
       18
```

Steve also bought ¼ pound of lunch meat. Lunch meat costs $3.96 per pound. To find the cost for ¼ pound, Steve divided by 4—the bottom number of the fraction ¼.

```
       $ .99  price per ¼ pound
  4 | $3.96  price per pound
       3 6
       36
       36
```

Figure the cost of each of these things. (Use the Common Fractions on page 128 to help you see the amount of each fraction.)

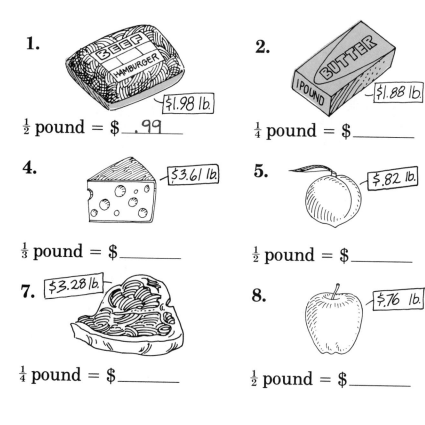

1. BEEF HAMBURGER $1.98 lb.
$\frac{1}{2}$ pound = $ .99

2. BUTTER 1 POUND $1.88 lb.
$\frac{1}{4}$ pound = $_____

3. $.36 lb.
$\frac{1}{2}$ pound = $_____

4. $3.61 lb.
$\frac{1}{3}$ pound = $_____

5. $.82 lb.
$\frac{1}{2}$ pound = $_____

6. $3.52 lb.
$\frac{1}{2}$ pound = $_____

7. $3.28 lb.
$\frac{1}{4}$ pound = $_____

8. $.76 lb.
$\frac{1}{2}$ pound = $_____

9. BEST COFFEE BEANS $6.48 lb.
$\frac{1}{8}$ pound = $_____

10.

¼ pound = $ _____

11.

⅓ pound = $ _____

12.

¼ pound = $ _____

13.

½ pound = $ _____

14.

¼ pound = $ _____

15.

⅛ pound = $ _____

16.

½ pound = $ _____

17.

¼ pound = $ _____

18.

⅓ pound = $ _____

19.

¼ pound = $ _____

20.

⅛ pound = $ _____

21.

¼ pound = $ _____

 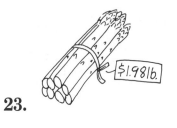

22.

⅓ pound = $ _____

23.

½ pound = $ _____

24.

⅓ pound = $ _____

Buying Material

Gant's Department Store was having a sale on material.
It sold material by the yard, but many people bought
less than a yard.

One kind of material cost $3.76 per
yard. Mary Rossi, a friend of Carla's,
bought ¾ yard to make a shirt. How
much did it cost? To find out, first
divide the cost per yard by 4. Your
answer is the cost of ¼ yard.

$$
\begin{array}{r}
\$\ .94 \\
4\,\overline{)\$3.76} \\
3\,6 \\
\hline
16 \\
16 \\
\hline
\end{array}
$$

cost of $\frac{1}{4}$ yard
cost per yard

Next multiply the cost of ¼ yard by
3—the top number in the fraction ¾.
Your answer is the cost of ¾ yard.

$$
\begin{array}{r}
\$\ .94 \\
\times\ 3 \\
\hline
\$2.82 \\
\end{array}
$$

cost of $\frac{1}{4}$ yard

cost of $\frac{3}{4}$ yard

Find the cost of ¾ yard of each of these materials.

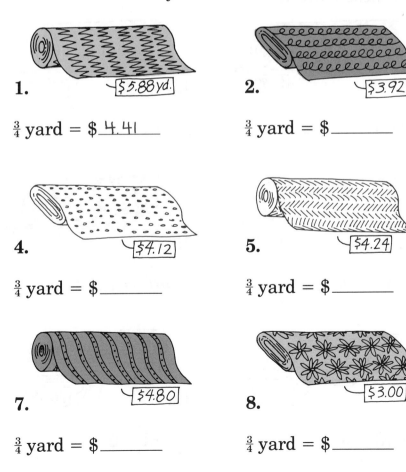
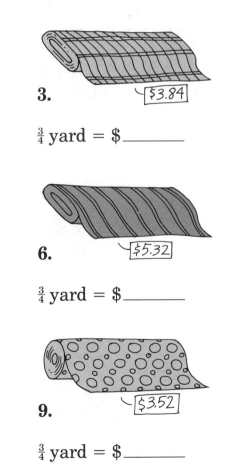

1. $5.88 yd.

$\frac{3}{4}$ yard = $ 4.41

2. $3.92

$\frac{3}{4}$ yard = $_____

3. $3.84

$\frac{3}{4}$ yard = $_____

4. $4.12

$\frac{3}{4}$ yard = $_____

5. $4.24

$\frac{3}{4}$ yard = $_____

6. $5.32

$\frac{3}{4}$ yard = $_____

7. $4.80

$\frac{3}{4}$ yard = $_____

8. $3.00

$\frac{3}{4}$ yard = $_____

9. $3.52

$\frac{3}{4}$ yard = $_____

Another kind of material cost $5.61 per yard. Mary bought $\frac{2}{3}$ yard. How much did it cost? To find out, first divide the cost per yard by 3. Your answer is the cost of $\frac{1}{3}$ yard.

$$
\begin{array}{r}
\$1.87 \quad \text{price for } \tfrac{1}{3} \text{ yard} \\
3\overline{)\$5.61} \quad \text{price per yard} \\
\underline{3} \\
2\,6 \\
\underline{2\,4} \\
21 \\
\underline{21}
\end{array}
$$

Next, multiply the cost of $\frac{1}{3}$ yard by 2. Your answer is the cost of $\frac{2}{3}$ yard.

$$
\begin{array}{r}
\$1.87 \quad \text{cost per } \tfrac{1}{3} \text{ yard} \\
\underline{\times\ 2} \\
\$3.74 \quad \text{cost of } \tfrac{2}{3} \text{ yard}
\end{array}
$$

Find the cost of $\frac{2}{3}$ yard of each of these materials.

10. $4.41 Yd.

$\frac{2}{3}$ yard = $ __2.94__

11. $7.89

$\frac{2}{3}$ yard = $_____

12. $9.00

$\frac{2}{3}$ yard = $_____

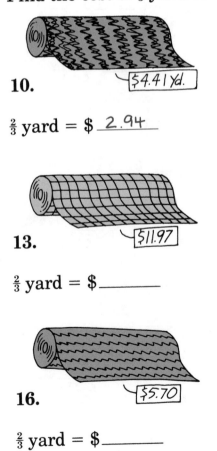

13. $11.97

$\frac{2}{3}$ yard = $_____

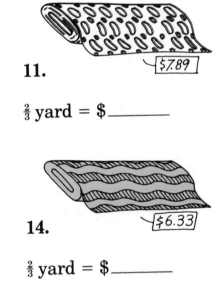

14. $6.33

$\frac{2}{3}$ yard = $_____

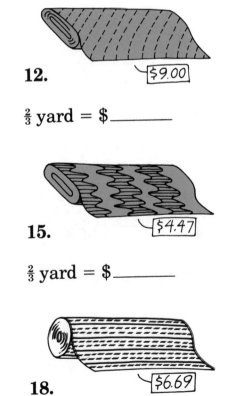

15. $4.47

$\frac{2}{3}$ yard = $_____

16. $5.70

$\frac{2}{3}$ yard = $_____

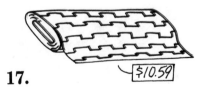

17. $10.59

$\frac{2}{3}$ yard = $_____

18. $6.69

$\frac{2}{3}$ yard = $_____

Shopping Quiz

Figure the cost of each of these things.
Write your answers in the blanks.

 $.48 lb.

1.

$\frac{3}{8}$ lb. = $ ___ .18 ___
(pound)

 $.55 lb.

2.

$\frac{3}{5}$ lb. = $ _____

 $4.36 yard

3.

$\frac{3}{4}$ yard = $ _____

 $.72 lb.

4.

$\frac{5}{8}$ lb. = $ _____

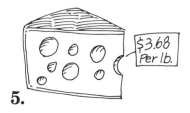 $3.68 Per lb.

5.

$\frac{3}{8}$ lb. = $ _____

 $3.95 Per lb.

6.

$\frac{4}{5}$ lb. = $ _____

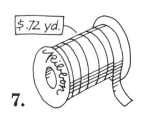 $.72 yd.

7.

$\frac{5}{6}$ yard = $ _____

 $1.28 lb.

8.

$\frac{7}{8}$ lb. = $ _____

 $.95 lb.

9.

$\frac{4}{5}$ lb. = $ _____

 $1.04 lb.

10.

$\frac{3}{8}$ lb. = $ _____

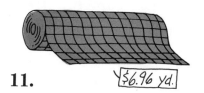 $6.96 yd.

11.

$\frac{7}{8}$ yard = $ _____

 $.96 lb

12.

$\frac{3}{4}$ lb. = $ _____

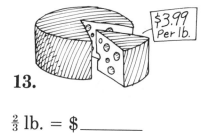 $3.99 Per lb.

13.

$\frac{2}{3}$ lb. = $ _____

 $.84 lb.

14.

$\frac{5}{6}$ lb. = $ _____

 $1.84 lb.

15.

$\frac{5}{8}$ lb. = $ _____

Grocery Quiz

Figure the cost of each of these things.
Write your answers in the blanks.

1.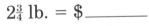

$2\frac{3}{4}$ lb. = $ _____

2.

$2\frac{2}{3}$ lb. = $ _____

3.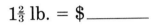

$1\frac{2}{3}$ lb. = $ _____

4.

$2\frac{3}{8}$ lb. = $ _____

5.

$1\frac{7}{8}$ lb. = $ _____

6.

$3\frac{3}{4}$ lb. = $ _____

7.

$2\frac{1}{4}$ lb. = $ _____

8.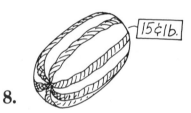

$6\frac{3}{5}$ lb. = $ _____

9.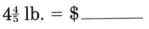

$4\frac{4}{5}$ lb. = $ _____

10.

$3\frac{1}{8}$ lb. = $ _____

11.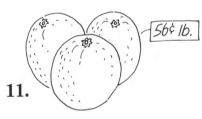

$2\frac{3}{4}$ lb. = $ _____

12.

$9\frac{2}{3}$ lb. = $ _____

13.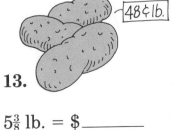

$5\frac{3}{8}$ lb. = $ _____

14.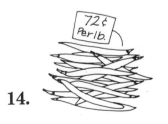

$2\frac{5}{8}$ lb. = $ _____

15.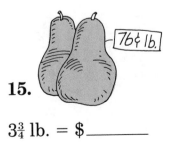

$3\frac{3}{4}$ lb. = $ _____

Tomatoes for Sale

Steve Pollard needed to buy tomatoes for The Sandwich House. Farmer's Yard was selling tomatoes for 64¢ a pound. Figure the cost of each basket below. Write your answers in the blanks.

1.

$3\frac{1}{2}$ pounds = $\underline{2.24}$

2.

$5\frac{3}{4}$ pounds = $\underline{\hspace{1cm}}$

3.

$9\frac{1}{4}$ pounds = $\underline{\hspace{1cm}}$

4.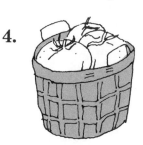

$12\frac{3}{8}$ pounds = $\underline{\hspace{1cm}}$

5.

$14\frac{1}{2}$ pounds = $\underline{\hspace{1cm}}$

6.

$15\frac{5}{8}$ pounds = $\underline{\hspace{1cm}}$

7.

$17\frac{1}{4}$ pounds = $\underline{\hspace{1cm}}$

8.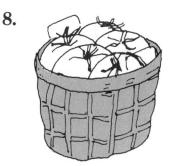

$20\frac{3}{4}$ pounds = $\underline{\hspace{1cm}}$

9.

$21\frac{3}{8}$ pounds = $\underline{\hspace{1cm}}$

10.

$24\frac{1}{2}$ pounds = $\underline{\hspace{1cm}}$

Money Problems

Count the money on the left side of the page. Write the total amount in the first blank on the right side of the page. Then find the fraction of that amount asked for. Write that amount in the second blank.

1.

Total amount = $_____

$\frac{3}{4}$ of amount = $_____

2.

Total amount = $_____

$\frac{2}{3}$ of amount = $_____

3.

Total amount = $_____

$\frac{7}{8}$ of amount = $_____

4.

Total amount = $_____

$\frac{2}{5}$ of amount = $_____

Working Overtime

In June, Carla Williams started working full time for Runaway Sports Shop. Mr. Swift was very pleased with Carla's work, so he gave her a raise. Carla now earned $4.90 an hour. She worked 40 hours a week. But sometimes Mr. Swift asked Carla to work more than 40 hours a week. This is called working *overtime*.

For every hour over 40 that Carla worked in a week, Mr. Swift paid her *time and a half.* Time and a half means that for every hour Carla worked overtime, she earned her regular hourly wage plus half that amount.

Here is how Carla figured how much she earned an hour at time and a half.

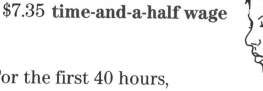

$$
\begin{array}{r}
\underline{\$2.45} \text{ half of hourly wage}\\
2\ \overline{)\ \$4.90}\ \text{ hourly wage}
\end{array}
$$

$$
\begin{array}{r}
\$4.90 \text{ hourly wage}\\
+\ \underline{\$2.45} \text{ half of hourly wage}\\
\$7.35 \text{ time-and-a-half wage}
\end{array}
$$

One week Carla worked 46 hours. For the first 40 hours, she earned her regular hourly wage. For the last 6 hours, she earned time and a half. Here is how Carla figured how much she earned for the week.

$$
\begin{array}{r}
\$4.90 \text{ hourly wage}\\
\underline{\times\ 40}\ \text{ hours worked}\\
\$196.00 \text{ regular wages}
\end{array}
$$

$$
\begin{array}{r}
\$7.35 \text{ time-and-a-half wage}\\
\underline{\times\ 6}\ \text{ overtime hours worked}\\
\$44.10 \text{ time-and-a-half wages}
\end{array}
$$

$$
\begin{array}{r}
\$196.00 \text{ regular wages}\\
+\ \underline{44.10}\ \text{ time-and-a-half wages}\\
\$240.10 \text{ total wages}
\end{array}
$$

Another week, Carla worked 43 hours. Answer these questions about how much she earned that week. Remember, Carla's regular hourly wage is $4.90.

1. How many hours did Carla work at her regular hourly wage? _____

2. How much did Carla earn at her regular wage? _____

3. How many hours overtime did Carla work? _____

4. How much did Carla earn at the time-and-a-half wage? _____

5. How much did Carla earn altogether for 43 hours' work? _____

One week, Runaway Sports Shop had a big sale. The store was very busy, so Carla worked overtime. That week, Carla worked 48 hours. Answer these questions about how much she earned that week.

6. How many hours did Carla work at her regular hourly wage? _____

7. How much did Carla earn at her regular wage? _____

8. How many hours overtime did Carla work? _____

9. How much did Carla earn at the time-and-a-half wage? _____

10. How much did Carla earn altogether for 48 hours' work? _____

NAME _____

Time and a Half

The hours Carla worked in six weeks are shown below.
Her regular wage is $4.90 an hour. She gets time and a
half for working overtime (more than 40 hours a week).
Fill in the blanks to show how much Carla earned each
week.

Week Ending	Hours Worked	Regular Earnings	Time-and-a-Half Earnings	Total Earnings
8/3	40			
8/10	42			
8/17	45			
8/24	38			
8/31	41			
9/7	44			

Answer these questions about Carla's earnings.

1. Did Carla get paid overtime every week? _____

2. For which weeks did Carla earn time and a half? _____

3. For which week did Carla earn the most money? _____

4. For which week did Carla earn the least money? _____

5. For which week did Carla earn the most overtime? _____

6. How many hours did Carla work overtime during the six weeks? _____

7. How much did she earn in time-and-a-half wages during the six weeks? _____

8. Carla deposited $\frac{3}{4}$ of her time-and-a-half earnings in her savings account. How much money did she deposit during the six weeks? _____

Time and a Half

The number of hours each of the employees of the Chocolate Factory worked one week is shown below. Anyone who worked overtime (more than 40 hours) gets time and a half for the overtime hours. Figure the time-and-a-half wage for each employee. Then fill in the blanks to show how much each employee earned for the week.

	Employee	Regular Hourly Wage	Time-and-a-Half Hourly Wage	Hours Worked	Regular Earnings	Time-and-a-Half Earnings	Total Earnings
1.	Tom Jay	$4.50		47			
2.	Kay Fox	$4.90		41			
3.	Ed Eng	$4.70		38½			
4.	Sue Orr	$4.30		43			
5.	Jim Day	$4.60		40			
6.	Roy Wax	$5.00		46			
7.	Joe Zim	$4.40		42			
8.	Ida Lee	$5.10		44			
9.	Al Sax	$4.80		45			
10.	Fay Tut	$6.00		40			
11.	Ben Gee	$5.30		48			
12.	Sal Ott	$5.50		49			
13.	Ike Foy	$6.20		39½			
14.	Ann Bly	$5.30		47			
15.	Pat Uno	$5.60		50			

Benefits

When Carla started working full time, she began to get *benefits*. This means Runaway Sports Shop paid for part of her health insurance costs. They would also pay for Carla's sick leave and vacation time.

With health insurance, Carla had to pay only a part of her doctor and dentist bills. The insurance plan had a *deductible* of $100. This means Carla would have to pay the first $100 of her bills each calendar year. After that, the insurance company would pay 80% of each bill.

To pay for her part of the insurance, another deduction was taken from Carla's paycheck. Her paycheck now showed these deductions:

REGULAR WAGES $424.67	OVERTIME WAGES $88.20	TOTAL GROSS WAGES $512.87	FEDERAL INCOME TAX $58.00	STATE INCOME TAX $6.40

	FICA $38.52	SDI $6.15	HEALTH INSURANCE $28.00	NET WAGES $375.80

Carla gets paid twice a month. This means she gets 24 paychecks a year. How much does her health insurance cost her for one month? _____ For one year? _____

Here are the amounts of Carla's doctor and dentist bills for one year. For each bill, figure out how much Carla will pay and how much the insurance company will pay. Write your answers in the blanks.

Bills	Carla pays	Insurance pays
$100 (dentist)	$100 (deductible)	0
	Carla pays 20%	Insurance pays 80%
$ 70 (dentist)	$ 14.00	_____
$ 45 (doctor)	_____	_____
$ 15 (medicine)	_____	_____

When Carla worked part time, she did not get paid if she was sick and missed work. Now that she worked full time, she could miss some days and still get paid.

Runaway Sports Shop gave full-time employees one-half day of sick leave per month. Carla would earn six days of sick leave each year. She could not build up sick leave from year to year. If she was sick more than six days in any year, she would only get paid for six of those days.

Figure the amount of sick leave Carla will earn. Write your answers in the blanks.

Time worked	Sick leave earned
1 month	½ day
2 months	_____
3 months	_____
6 months	_____
10 months	_____

Carla also began to earn vacation time. She would earn two weeks vacation each year. After she had worked full time for five years, she would earn three weeks a year.

Carla would have to work full time for six months before she could use any of her vacation. Again, she could not build up vacation time from year to year. If she did not use all of her vacation days within a year, she would lose the days she didn't use.

Figure the amount of vacation Carla will earn. Write your answers in the blanks.

Time worked	Vacation earned
2 years	2 weeks
6 months	1 week
9 months	_____
6 months	_____
1 year	_____

Saving Money

Carla was happy to work overtime at Runaway Sports Shop. She liked her job and earning the overtime money. She was saving her money for a vacation. She planned to visit her sister in New Orleans in February. Her goal was to save $500.00 to pay for the trip. She had six months to save up the money.

To help make saving money fun, Carla drew a thermometer to show how much money she saved. Each time she deposited money in her savings account, she would color in part of the thermometer. When the whole thermometer was colored in, she would have reached her goal. Carla would have saved 100% of the money she needed for her trip. 100% is the whole amount—$500.00.

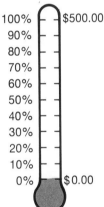

Carla looked in her savings passbook to see how much money she had already saved. She had 10% of her goal. How much money did she have in her savings account?

To find out, change 10% to a decimal fraction—.10. Then multiply the total amount of money to be saved ($500.00) by .10. Move the decimal point four places to the left in your answer. 10% of $500.00 is $50.00.

$$10\% = .10$$

$500.00	**total amount**
× .10	
$50.0000	**10% of total**

Figure the following percent amounts that Carla saved.
Write each amount next to the thermometer.

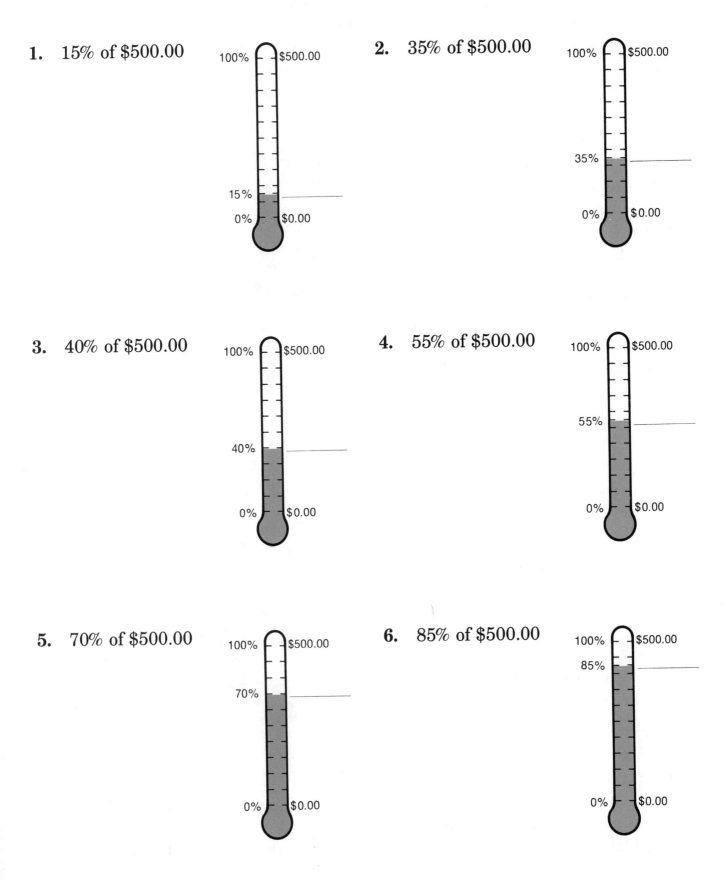

1. 15% of $500.00

2. 35% of $500.00

3. 40% of $500.00

4. 55% of $500.00

5. 70% of $500.00

6. 85% of $500.00

Figuring Percent

When Carla began working full time, she was paid twice a month. Figure the following percent amounts that Carla saved from every paycheck for six months. Round off your answers to the nearest cent. Then answer the last two questions.

1. 12% of $229.10 = $27.49

$$
\begin{array}{r}
\$\ 229.10 \\
\times\ \ \ .12 \\
\hline
45820 \\
22910\ \ \\
\hline
\$27.4920
\end{array}
$$

2. 13% of $312.00 = _____

3. 18% of $298.26 = _____

4. 16% of $284.10 = _____

5. 17% of $273.36 = _____

6. 14% of $277.13 = _____

7. 13% of $318.12 = _____

8. 19% of $294.24 = _____

9. 10% of $289.43 = _____

10. 15% of $305.24 = _____

11. 16% of $327.66 = _____

12. 14% of $342.89 = _____

13. How much money did Carla save in six months? _____

14. Did Carla reach her goal of saving $500.00 in six months? _____

Figuring Raises

In June, Steve Pollard started working full time for The Sandwich House. The Sandwich House was doing very well, so the owner gave all the employees raises. Figure the amount of each raise and each employee's new earnings.

	Employee	Regular Monthly Earnings	Percent Raise	Amount of Raise	Total New Earnings
1.	Steve Pollard	$800.00	18%		
2.	Susan Jones	$830.00	15%		
3.	David Wolfe	$775.00	16%		
4.	Ruth Royer	$850.00	12%		
5.	Kim Lee	$840.00	14%		
6.	Mike Bowen	$790.00	20%		
7.	Jose Vega	$875.00	15%		
8.	Mary Wang	$1,000.00	10%		
9.	Nina Perez	$945.00	13%		
10.	John Conner	$870.00	11%		

Steve's Circle Graph

Like Carla, Steve had been saving part of his earnings. He had saved enough money for a down payment on a used car. But after he bought the car, he had to make car payments every month. So he needed to be very careful about spending his money.

To help keep track of his spending, Steve drew a circle graph. The graph showed what percent of his take-home pay he spent on different things for one month. Steve's total take-home pay for that month was $1,100.00 (wages, overtime pay, and tips).

28% car
25% rent
5% spending money
14% food
10% savings
6% clothes
6% health care
4% personal needs
2% gas and electricity

Use the circle graph to answer the questions below. Use $1,100.00 as 100% of Steve's monthly earnings.

1. Steve rented an apartment with a friend. How much money did he spend on rent? _____

2. How much money did Steve spend on food? _____

3. How much money did he spend on gas and electricity? _____

4. How much money did Steve spend on personal needs, such as laundry and things from the drugstore? _____

5. How much did he spend on health care for doctor and dentist bills and medicine? _____

6. How much money did Steve spend on clothes? _____

7. How much money did he spend on his car for car payments, gas, and car insurance? _____

8. How much money did Steve deposit in his savings account? _____

9. How much money did Steve have left over for spending money? _____

Steve's Spending

Steve did not always earn $1,100.00 a month. Sometimes he did not work overtime. And the amount of money he earned in tips was different every month.

One month, Steve earned $1,000.00 in take-home pay. So he had to cut down on his spending. Steve could not cut down on some things. He still had to pay $280 for rent. He had to pay $165 for his car payment and $80 for his car insurance. So Steve had to cut his spending on other things.

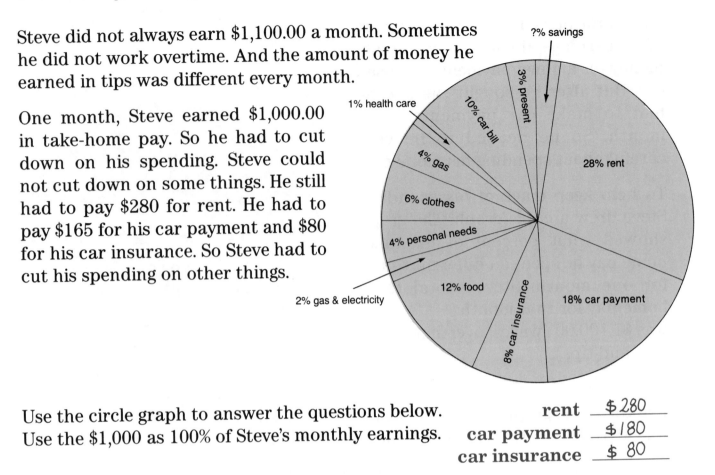

?% savings
1% health care
3% present
10% car bill
4% gas
6% clothes
4% personal needs
28% rent
2% gas & electricity
12% food
8% car insurance
18% car payment

Use the circle graph to answer the questions below.
Use the $1,000 as 100% of Steve's monthly earnings.

rent	$280
car payment	$180
car insurance	$80

1. Steve cut down on his food bill by not eating out.
   How much money did he spend on food? _____

2. He used less electricity that month.
   How much money did Steve spend on gas and electricity? _____

3. Steve spent a little less on personal needs.
   How much money did he spend on personal needs? _____

4. Steve bought only the clothes he really needed.
   How much money did he spend on clothes? _____

5. He saved money on gas by driving less.
   How much money did Steve spend on gas? _____

6. Steve had to buy some medicine.
   How much money did he spend on health care? _____

7. Steve's car broke down and he had to pay to fix it.
   How much money did he spend for his car bill? _____

8. Steve had to buy a present for a good friend.
   How much money did he spend on the present? _____

9. How much money did Steve have left for his savings account? _____

Earning Commission

Carla Williams went to Travelon Tours to book her trip to New Orleans. Her friend, Tom Otani, worked there. Tom told Carla how Travelon Tours earned its money.

For every airline ticket Travelon Tours sold, the airline paid them 5% of the cost of the ticket.

For every hotel booking they made, the hotel paid them 10% of the cost of the room.

For every cruise they sold, the cruise company paid them 15% of the cost of the cruise.

The money Travelon Tours earned is called *commission*. Commission is money earned from selling.

The map shows the cost of airline tickets between different cities. The ticket prices are for *one way*. Use the map to figure Travelon Tour's commission on the following bookings.

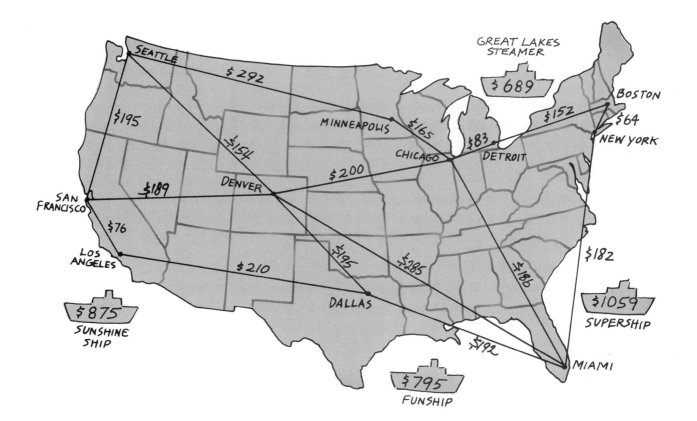

Trip	Bookings	Cost	Commission	Total Commission
1. AIR	San Francisco to Los Angeles	$ _76.00_	$ _3.80_	
	Los Angeles to San Francisco	$ _76.00_	$ _3.80_	$ _7.60_
2. AIR	Minneapolis to Seattle	$_____	$_____	
	Seattle to Minneapolis	$_____	$_____	
HOTEL	$46 a night for 4 nights	$_____	$_____	$_____
3. AIR	Chicago to Miami	$_____	$_____	
	Miami to Denver	$_____	$_____	
	Denver to Chicago	$_____	$_____	$_____
4. AIR	Boston to Detroit	$_____	$_____	
	Detroit to Chicago	$_____	$_____	
CRUISE	Great Lakes Steamer	$_____	$_____	$_____
5. AIR	Los Angeles to Dallas	$_____	$_____	
	Dallas to Miami	$_____	$_____	
HOTEL	$59 a night for 7 nights	$_____	$_____	$_____
6. AIR	Boston to New York	$_____	$_____	
	New York to Miami	$_____	$_____	
CRUISE	Supership Cruise	$_____	$_____	$_____
7. AIR	Denver to Seattle	$_____	$_____	
	Seattle to San Francisco	$_____	$_____	
	San Francisco to Denver	$_____	$_____	
HOTEL	$55 a night for 5 nights	$_____	$_____	$_____
8. AIR	Dallas to Los Angeles	$_____	$_____	
	Los Angeles to Dallas	$_____	$_____	
HOTEL	$45 a night for 2 nights	$_____	$_____	
CRUISE	Sunshine Cruise	$_____	$_____	$_____
9. AIR	Seattle to Minneapolis	$_____	$_____	
	Minneapolis to Chicago	$_____	$_____	
	Chicago to Denver	$_____	$_____	
	Denver to Seattle	$_____	$_____	
HOTEL	$39 a night for 12 nights	$_____	$_____	$_____
10. AIR	Denver to Dallas	$_____	$_____	
	Dallas to Miami	$_____	$_____	
	Miami to Denver	$_____	$_____	
HOTEL	$47 a night for 10 nights	$_____	$_____	
CRUISE	Funship Cruise	$_____	$_____	$_____

Carla's Trip

Late in February, Carla went to Travelon Tours to pay for her trip to New Orleans. Tom Otani gave her her airline ticket. Her ticket showed the time of her flights.

On her flight to New Orleans, Carla must change planes in St. Louis.

On her flight to Chicago, Carla must change planes in Memphis.

Travelon Tours	
FLIGHT TIMES	
February 23	
Leave Chicago	7:30 A.M.
Arrive St. Louis	8:30 A.M.
--CHANGE PLANES--	
Leave St. Louis	9:10 A.M.
Arrive New Orleans	11:10 A.M.
March 1	
Leave New Orleans	6:50 P.M.
Arrive Memphis	7:50 P.M.
--CHANGE PLANES--	
Leave Memphis	8:40 P.M.
Arrive Chicago	10:10 P.M.

Study the list of Carla's flight times.
Then answer the questions below.

1. How long is the flight from Chicago to St. Louis?   _____

2. How long is the flight from St. Louis to New Orleans?   _____

3. What is the total flying time from Chicago to New Orleans?   _____

4. How much time does Carla have in St. Louis to change planes?   _____

5. On the return trip, how much time does Carla have in Memphis to change planes?   _____

6. What is the total flying time from New Orleans to Chicago?   _____

7. What is the total flying time for the round trip?   _____

Carla paid for her trip with a check. The cost of the airline ticket for the round trip was $328.00. She had to pay 5% sales tax.

Figure the total cost of Carla's ticket. Then fill out the check below to Travelon Tours for the total amount. Use today's date and sign your name. Then answer the questions below.

UNITED NATIONAL BANK
Hometown, Illinois

NUMBER 348

09-38
241

_____ 19 _____

PAY TO
THE ORDER OF _____ $ _____

_____ DOLLARS
nonnegotiable

⑈0241⑈ 0938⑈ 618⑈51⑈06

8. Carla had saved $500.00 for her trip. She planned to spend ¾ of her savings on her ticket. How much less than that did she spend?

_____

9. What is the one-way cost of the trip?

_____

10. Carla's flight from New Orleans to Chicago covered 840 miles. What was the cost per mile?

_____

11. In all, the trip was 1702 air miles. It took 5.5 hours of flying time. How many miles were covered each hour? (Round off your answer to the nearest mile.)

_____

12. In New Orleans, Carla took a taxi to her sister's house. The ride was 13.4 miles and cost $18.50. What was the cost per mile?

_____

The Weekend Trip

Carla and her sister, Maria, made a weekend trip. They rented a car and drove along the Mississippi River. They spent one night in a hotel. They each paid half the total cost of the trip.

Answer the questions to find out how much the trip cost.

1. Dixie Rent-a-Car charged $38.95 a day for a car. Mileage was free. How much would it cost to rent a car for two days? _____

2. Travelcar charged $24.75 a day and $.25 a mile. The trip was 200 miles. How much would it cost to rent the car for two days? _____

3. The sisters rented a car from the company with the lower price. Which company was that? _____

4. Carla and Maria drove 115 miles and used 4.6 gallons of gas. How many miles per gallon did the car get? _____

5. Carla and Maria stopped for lunch. Add up their bill on the right.

   They also left a tip that was 15% of the total. How much was the tip?

   _____

RIVERBOAT CAFE	
2 fish specials ($5.95 each)	
2 soft drinks ($.85 each)	
2 pieces of pie ($1.95 each)	
2 coffees ($.85 each)	
Subtotal	
Add 5% tax	
THANK YOU! TOTAL	

6. The hotel cost $56.00 a night. They had to pay 5% tax on the room. How much did the hotel cost altogether? _____

7. That night, the sisters spent $28.75 for dinner. The next day, they spent $12.25 for lunch. For the whole trip, they spent $16.00 for gas. How much did these things cost altogether? _____

8. How much did the sisters spend for the whole trip? _____

9. How much did Carla pay for $\frac{1}{2}$ the total cost of the trip? _____

10. Carla started out with $500.00. She spent $344.40 for her airline ticket and $18.50 for the taxi ride. How much money did she have left after her weekend trip?

Vacation Maze

Take your own trip! Start at the top left square of the maze. Follow the openings through the squares, subtracting the money you spend as you go. (Do your work on another piece of paper.) Then fill in the blank at the end of the maze.

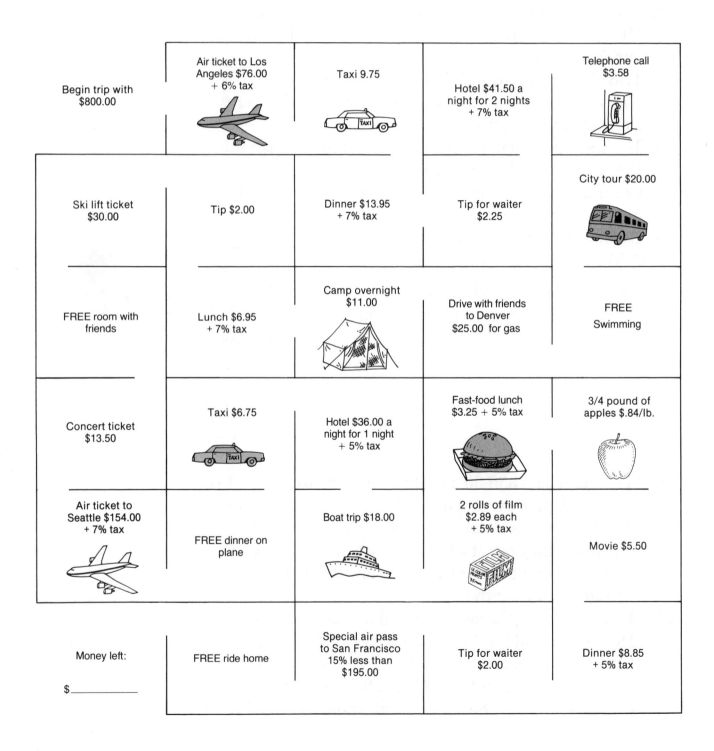

Begin trip with $800.00

Air ticket to Los Angeles $76.00 + 6% tax

Taxi 9.75

Hotel $41.50 a night for 2 nights + 7% tax

Telephone call $3.58

Ski lift ticket $30.00

Tip $2.00

Dinner $13.95 + 7% tax

Tip for waiter $2.25

City tour $20.00

FREE room with friends

Lunch $6.95 + 7% tax

Camp overnight $11.00

Drive with friends to Denver $25.00 for gas

FREE Swimming

Concert ticket $13.50

Taxi $6.75

Hotel $36.00 a night for 1 night + 5% tax

Fast-food lunch $3.25 + 5% tax

3/4 pound of apples $.84/lb.

Air ticket to Seattle $154.00 + 7% tax

FREE dinner on plane

Boat trip $18.00

2 rolls of film $2.89 each + 5% tax

Movie $5.50

Money left:

$_____

FREE ride home

Special air pass to San Francisco 15% less than $195.00

Tip for waiter $2.00

Dinner $8.85 + 5% tax

Posttest I

Answer these questions.

1. Suppose you work from 3:30 P.M. to 9:00 P.M. each day, Monday through Friday. You take an unpaid break from 6:00 to 6:30. How many hours do you work each week? _____

2. You work in a restaurant. Your hourly wage is $4.40. One evening you work from 5:30 to 9:30 and make $34.20 altogether. How much do you make in tips that evening? _____

3. You just got your first paycheck. The gross amount was $187.15, and the net amount was $150.55. The state tax, FICA, and SDI totaled $17.60. How much was the federal income tax? _____

4. You buy several tools at a hardware store. The subtotal is $42.50, and the tax is 6%. What is the total cost? _____

5. Your take-home pay is $846.00 per month. You spend 1/6 of it on your car payment. How much is your car payment? _____

6. You are collecting money from 14 people at work. You want to buy a co-worker a present. You need to collect $37.50. If all 15 of you give the same amount, how much will each person need to give? _____

7. You buy $2\frac{5}{8}$ pounds of salmon at $5.60 a pound. How much do you spend on salmon. _____

8. Suppose your company gives its employees eight days of sick leave each year. You have worked at the company for nine months. How many days of sick leave have you earned? _____

9. You are depositing a ten-dollar bill, a one-dollar bill, three quarters, and a check made out to you for $16.92. The bank number on the check is 12–36. Fill out the deposit slip, using today's date.

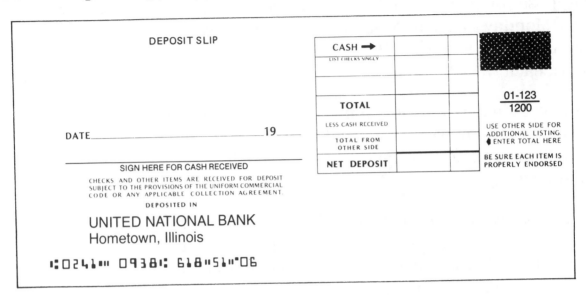

10. Write each of these entries in the correct column of the passbook below. Figure the new balance after each entry.

Date	Entry	Amount
9/7	Deposit	125.00
9/14	Deposit	50.00
9/15	Interest	(5% of 175.00)
9/28	Withdrawal	30.00
10/12	Deposit	85.00
10/19	Withdrawal	27.00

DATE	WITHDRAWAL	DEPOSIT	INTEREST	BALANCE

Posttest II

Use the arithmetic skills you have learned to do these
problems. Remember to put the decimal point in the
right place in your answers.

1.
```
   25
   86
   47
 + 31
 ----
  189
```

2.
```
    40
    98
   102
 + 158
 -----
```

3.
```
   986
 - 109
 -----
```

4.
```
 $12.46
 ×    3
 ------
```

5.
```
 $8.67
 - 4.35
 ------
```

6.
```
 $48.89
 ×    9
 ------
```

7.
```
  $1.35
   1.48
   9.10
 + 6.55
 ------
```

8.
```
 $10.58
   9.64
 + 8.12
 ------
```

9. 2) $5.72

10.
```
 $86.32
 - 78.30
 ------
```

11. 8) $9.84

12.
```
 $60.40
 - 53.25
 ------
```

13.
```
    47
    53
   188
 + 243
 -----
```

14.
```
 $4.75
 ×  48
 -----
```

15.
```
  $1.50
   2.35
   4.10
 + 6.54
 ------
```

16.
```
 $85.46
 - 64.58
 ------
```

17. 9) $23.85

18.
```
 $258.63
 + 159.87
 -------
```

19.
```
 $45.82
  75.04
  58.67
 + 63.88
 ------
```

20.
```
 $126.38
 - 83.92
 -------
```

21.
```
 $238.11
 -  8.99
 -------
```

22.
```
 $7.99
 ×  67
 -----
```

23. 6) $218.64

24.
```
 $5.62
 ×  52
 -----
```

25. 7.5) 96

26. $356.80
25.71
+ 9.34

27. 13)$99.19

28. $100.00
× .01

29. $458.31
− 258.22

30. $376.80
× .05

31. 15)$131.10

32. $332.06
− 306.07

33. $479.00
× .29

34. $493.09
122.47
+ 258.58

35. 2.5)21.5

36. $208.86
− 180.62

37. $123.50
× .18

38. 7.7)34.65

39. $248.50
× .64

40. 11.8)749.3

41. Find 75% of $14.52. ___$ 10.89___

42. Find 42% of $38.50. _____

43. Find 8% of $17.25._____

44. Find 10% of $219.00. _____

45. Find 12% of $52.75. _____

46. Find $\frac{1}{2}$ of 980 lb. _____

47. Find $\frac{3}{4}$ of $663.24. _____

48. Find $\frac{5}{8}$ of 152 lb. _____

49. Find $\frac{2}{3}$ of $115.83. _____

50. Find $\frac{3}{8}$ of $29.44._____

Multiplication Table

Use this multiplication table to help you find the answers to multiplication problems.

×	1	2	3	4	5	6	7	8	9	10	11	12
1	1	2	3	4	5	6	7	8	9	10	11	12
2	2	4	6	8	10	12	14	16	18	20	22	24
3	3	6	9	12	15	18	21	24	27	30	33	36
4	4	8	12	16	20	24	28	32	36	40	44	48
5	5	10	15	20	25	30	35	40	45	50	55	60
6	6	12	18	24	30	36	42	48	54	60	66	72
7	7	14	21	28	35	42	49	56	63	70	77	84
8	8	16	24	32	40	48	56	64	72	80	88	96
9	9	18	27	36	45	54	63	72	81	90	99	108
10	10	20	30	40	50	60	70	80	90	100	110	120
11	11	22	33	44	55	66	77	88	99	110	121	132
12	12	24	36	48	60	72	84	96	108	120	132	144

Common Fractions

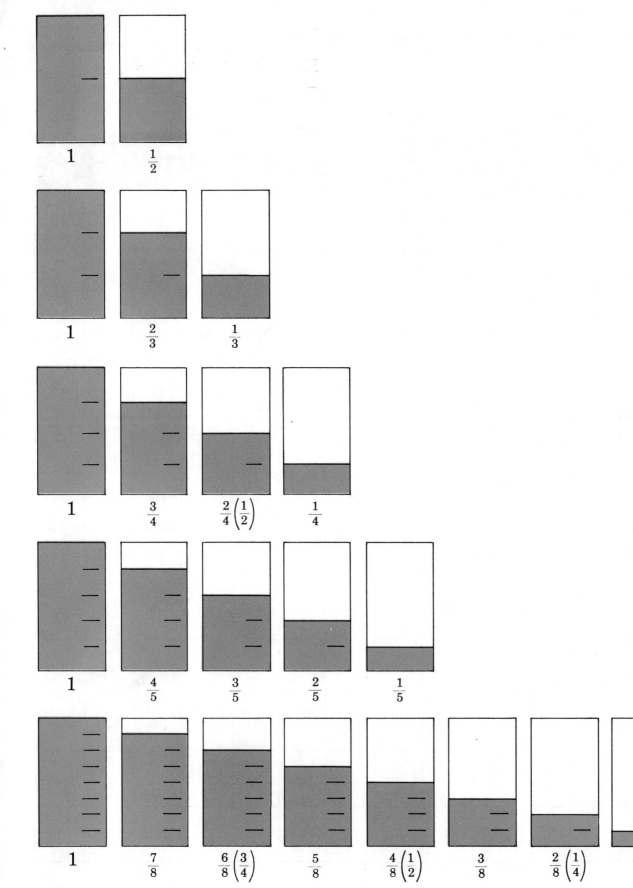